DURIA ANTIGUIOR, OR ANCIENT DORSETSHIRE. (See page 36.)

CURIOSITIES

OF

NATURAL HISTORY.

Second Series.

BY

FRANCIS T. BUCKLAND, M. A.,

STUDENT OF CHRIST CHURCH, OXFORD; ASSISTANT SURGEON SECOND LIFE GUARDS; AUTHOR OF CURIOSITIES OF NATURAL HISTORY, FIRST SERIES, ETC.

NEW YORK:

RUDD & CARLETON, 130 GRAND STREET.

LONDON: RICHARD BENTLEY.

M DCCC LX.

PRINTED FROM THE AUTHOR'S EARLY SHEETS.

CONTENTS.

A SLIGHT TRIBUTE

TO THE MEMORY

OF

My dear Mother,

FOR WHOSE EARLY INSTRUCTION AND PARENTAL CARE

I OWE A DEBT OF GRATITUDE

I CAN NEVER REPAY.

F. T. B.

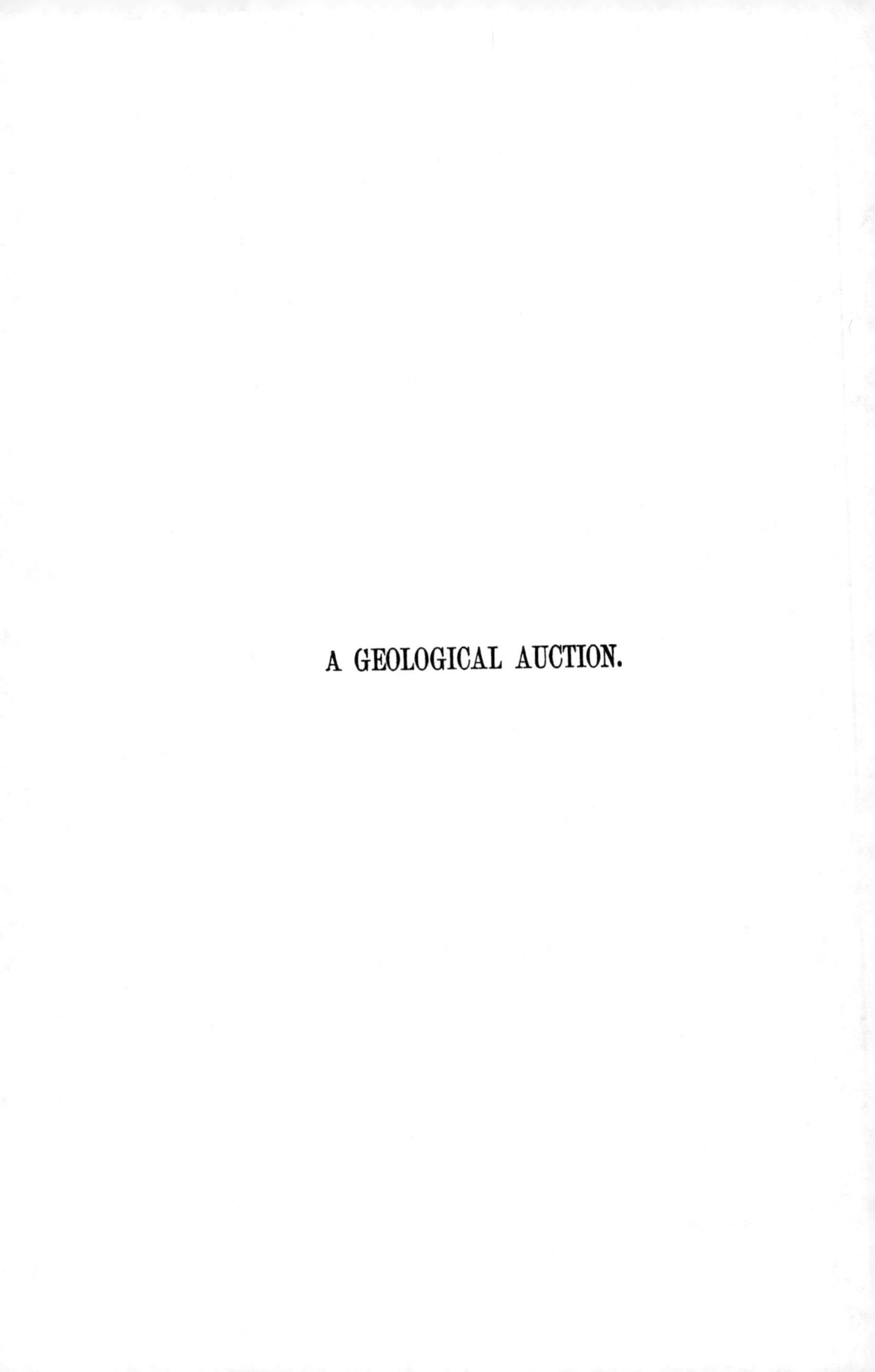

A GEOLOGICAL AUCTION.

PREFACE.

If we consider the numerous scientific and valuable works on Natural History which almost daily appear from the public press, it would seem almost superfluous to attempt to add another book on this subject. I have, nevertheless, made bold to publish this little work, because, in the first place, I do not think *too much* can be written about the ever-various and beautiful works of the great Creator; and, secondly, because there are many little things which in general are not thought worthy of being recorded, but which nevertheless have much interest to the true lover of Natural History.

In our leisure moments, when the business of the day is over, we can hardly walk along London streets

—and certainly not along the hedge-rows and fields of the country, or the wave-washed shore of the ocean—without finding almost at every step something or other worthy of observation; may-be our specimen is, and probably will be, common enough, yet, if it be rightly examined, it will be found to possess great interest, and to have an unwritten history of its own, which it should be our pleasure to interpret.

There is nothing so wearisome, or so destructive to the human mind, as the disease called "Nothing to do:" there is always and everywhere something to be done; there are no two places in this world exactly alike in their products, animal, vegetable, or mineral, and the objects you do *not* find in one place you *will* find in another.

If the eyes be instructed and trained to observe what is brought before their gaze, the mind is employed, and the feeling of weariness passes off: one fact follows another; a new observation may be tacked on to an old observation; the result being not only pleasure in discovery, but pleasure in recollecting and recording.

In the following pages I do not pretend to great things: they are more a collection of occasional notes than a literary production.

The *skeletons* of the "Gamekeeper's Museum" and of the "Hunt on the Sea-shore" have already appeared in that rising little periodical the *St. James's Medley;* in the Naturalist's columns of the *Field* newspaper; as well as in *Household Words;* and I here take this opportunity of thanking the editors of those publications for their kind permission to reproduce these articles in their present form. I have, however, added very much matter to them from subsequent notes and repeated observations.

To my friend Professor Quekett, Curator of the Museum of the Royal College of Surgeons, and other kind friends, I must return my sincere thanks for assistance during the progress of this little book.

Finally, the success of my First Series, which is now in its fifth thousand, has emboldened me to write a companion volume to it, and I trust the reader will be pleased to approve of this my second humble attempt to instruct and at the same time, if possible, to amuse.

The publication of my labors has been greatly delayed by the illness and, I regret to say, the death (May 4, 1860) of my lamented friend, T. Tardrew, Esq., Surgeon of the Regiment to which I have the honor to belong. My professional duties have, from this cause, for the last ten months, been heavier than usual, and it is only my spare time that I can devote to literary labors.

FRANCIS T. BUCKLAND,
2d Life Guards.

ATHENÆUM CLUB, PALL MALL,
June, 1860.

EXPLANATION OF THE FRONTISPIECE.

1 Ichthyosaurus Vulgaris.
2. Ichthyosaurus Tenuirostris (or the slender-jawed.)
3. Plesiosaurus Doliclodirus (or the Long-necked.)
4. Pterodactyles.
5. Lapedium Pablum (or the "Paving-stone" scaled fish.)
6. Pentacrinites Briaræus (or the Hundred-handed.)
7. Cuttle Fish.
8. Ammonites.
9. Crocodile.
10. Lamia or "Fossil Bird's nest.")

THE Frontispiece is an attempt to reproduce a remarkable drawing by my father's friend, the late lamented Sir Henry Delabeche. After lecturing from his professional chair in the University of Oxford, upon the numerous and valuable specimens of extinct animal life contained in his museum, it was his habit to distribute copies of Sir Henry's drawing, in order to bring to the minds of his audience the reality of the subjects on which he had been conversing. It is, I believe, one of the first attempts ever made to restore to their living forms these monster inhabitants of ancient seas, though this idea has been of late fully developed by Mr. Waterhouse Hawkins, in his admirable models in the gardens of the Crystal Palace.

The drawing by Sir H. Delabeche was called "Duria Antiguior," or Ancient Dorsetshire. For along the sea-girt coast of this county is found a vast charnel-house of the fossil bones of the monsters that must at one time have lived, preyed upon each other, and ultimately died, at or near this very spot, leaving their remains to be discovered and described by geologists, and their forms as in life to be again restored by the talented pencil of the great geologist above mentioned.

In the centre of the plate, at fig. 1, is seen the mighty Ichthyosaurus,* or the Lizard Fish; in form and structure not unlike the marine mammalia that play the same part at the present day as did these

* The remains of this animal are so abundantly found, that its osseous and dental structure is as well known to anatomists as that of a crocodile or other creature of our own time.

monsters in ancient oolitic times. And this is known, on account, first, of there being an entire absence of that peculiar modification of the bones which support the gills in fish; second, the presence of true bony nostrils and not olfactory bags placed in the skin unconnected with bone, as in fish; third, the articulation of the ribs to the spine, as in recent air-breathing animals. Ichthyosaurus had fins or paddles at its side, and a long tail, at the end of which according to Professor Owen's* recent discoveries, was a vertical, conical fleshy fin. It was a true, air-breathing animal, and could do what no whale or grampus of the present day is capable of accomplishing, viz.: the Ichthyosaurus could crawl upon the shore, and that most likely at periodical times as do the seal, walrus, &c. It had an enormous eyeball, which was larger in proportion to the skull than the eye in any other kind of animal; and this eye, having no eyelids, contained delicate humors, which, being liable to injury in a

* In the spring of this year, Professor Owen gave his course of Fullerian Lectures at the Royal Institution, Albemarle Street, in which he gave a full description of this and other animals. I have availed myself of my MS. notes, with his permission, in the following descriptions.

chopping sea, was composed of numerous thin and (probably) flexible bones, which encased the pupil. Owl-like, it probably pursued its prey at dusk of evening, by moonlight, or at early morning It had a formidable array of teeth, each of which was undermined by the gum of its successor, so that if a violent snap, or a too-vigorous-captured prey broke away the old tooth, the new one would come up in its place. In the engraving it is represented as making good use of these teeth, for it has caught and is about to devour Plesiosaurus, fig. 3.

This also is a curious whale-like creature, which has aptly been likened to " a turtle threaded through with the body of a snake." This animal was marine-aquatic in its habits ; but unlike the Ichthyosaurus, which was a deep-sea animal, it was a shore creature, and lived in the estuaries of brackish water ; and there, lurking under the oar-weed and other marine vegetation, obtained its prey by darting out its long neck, and seizing its prey with its sharp and formidable teeth, as is seen at fig. 4 (and also in the distance,) where an unfortunate Pterodactyle has not got out of the way quickly enough, and is suffering for his laziness; while his frightened companions are wheeling about in the air overhead, like frightened sea-gulls when one of their comrades has been captured or shot.

This Pterodactyle, or winged-fingered Saurian, was a monstrous beast, a true Saurian, but yet with leather-like wings like a bat: the only living approach to them is the insignificant little Draco Votaris of the Isles of the Indian Archipelago (see p. 11.)

At fig. 5 is seen a fish whose name is Lapedium, so called on account of its "pavement-like" scales; it has encountered in its peregrination an Ichthyosaurus who is making short work of him, and is about to gorge him down into its capacious stomach, in the same manner that a jack does a roach or dace. We know that Ichthyosaurus fed upon this fish, because its scales are found in the fossil Coprolithes. (See pp. 30-31.) In the Oxford Museum is the petrified stomach of an Ichthyosaurus that had died shortly after its dinner, as it had not had time to digest entirely the fish it had swallowed. The Ichthyosaurus, as seen in the engraving, did not refuse to eat cuttle-fish, and we know this because the ink of the cuttle-fish is found staining the fossil Coprolithes (see p. 7.)

Other fish whose remains are found fossil are seen swimming about in company with the young Ichthyosaurus, all enjoying life, and following the laws of nature which ordained that they should both prey upon one another, and be preyed upon themselves.

Sailing along the surface of the sea, upon which no human eye ever rested, are seen a fleet of the beautiful Ammonite shells (see p. 35.) Their remains are seen at the bottom of the sea (fig. 8,) where they would become gradually covered with mud, and converted into fossils, a theme for the geologists and an ornament of our cabinets.

At fig. 6 we see growing in great luxuriance a remarkable form of life—the Pentacrinite, or Stone Lily, so called on account of the pentangular, or five-sided shape of its supporting column. It consisted of innumerable calcareous joints, united by a fleshy material; it was, in fact, a "stalked star-fish," which is represented in existing seas by the Comatula, or Feather-stars of our own shores, and by the rare and all but extinct Pentacrinites of the West Indies.

For a full and beautifully illustrated description of the Pentacrinite, as well as the Ichthyosaurus, Plesiosaurus, Pterodactyle, and other creatures represented in the drawing, I must refer my readers to Dr. Buckland's "Bridgewater Treatise."

At fig. 10 is represented Lamia, or "Bird's-nest," of the Portland quarrymen, together with restorations of vegetation, the remains of which are found fossil in abundance.

At the bottom of this primæval sea are strewed the bones and carcasses of its inhabitants, both small and great. Saurians, fishes, mollusks and shells, have all yielded up their remains in obedience to the dictum which pronounces the sentence of death upon everything that has ever been animated with the breath of life.

In their unknown graves for thousands of past centuries, converted into hard marble like rock, they have lain; and hundreds of skeletons *will* lie till time is no more, leaving but a bare record of their former existence engraved in letters of stone on the shores which once formed the bed of an ancient ocean now long passed away.

Meanwhile let it be our privilege to read and inter pret the history of our planet as it existed when yet young in the starry firmament. Let us compare extinct forms of animal life with their modern living prototypes; and from the habits and instincts of animals around us learn, not only the laws which govern them as well as ourselves, the physiological causes which regulate *our* bodies as well as *their* bodies, but also endeavor to learn pleasurable lessons from daily scenes around us, and to withdraw the veil which frequently obscures the most enchanting scenes of nature from ordinary observation; and, above all, to join with the inspired writer when he admonishes us: "But ask now the Beasts and they shall teach thee, and the Fowls of the air and they shall tell thee; or speak to the Earth and it shall teach thee, and the Fishes of the sea shall declare unto thee!"

A GEOLOGICAL AUCTION.

A CATALOGUE *of the* COLLECTION OF MINERALS, FOSSILS *from various localities, Casts and Medallions, together with a variety of Miscellanies, the property of the late* VERY REV. DR. BUCKLAND, *Dean of Westminster.*—30th Jan., 1857.

IT is about three years ago that a sale took place at Stephens' Auction Rooms, in King Street, Covent Garden, of the private collection of minerals, fossils and curiosities collected by my father, the late Rev. Dr. Buckland, D. D., Dean of Westminster, &c. This great Geologist had been from his earliest days an enthusiastic collector; endowed with a wonderful acuteness of observation, he picked up and put away many things which a superficial observer would think worthless, but which he, with his usual eloquence and tact, would turn to account in some future lecture, as illustrating the former condition of the planet on which we live; and set forth, as one of the many instances of the wisdom and the goodness of the great Creator towards the most humble of his creatures.

The greater part of the collection which he had made during his numerous geological trips both in England and Europe, he bequeathed by will to the University of Oxford, in the following words:—

"I give and bequeath to the Vice-Chancellor of the University of Oxford, and his successors for the time being, for the use of my successors in the said University, as Readers in Geology for the time being, all my geological specimens, minerals, models, maps, and geological charts, drawings, sections and engravings connected with geology, which shall be in the Clarendon at the time of my decease."—*Extract from Dr. Buckland's Will.*

The University has but lately voted a sum of money to Dr. Buckland's accomplished successor in the Professorial chair, Professor Phillips, for the purpose of arranging and labelling the specimens, previous to their removal from the Clarendon Buildings, to the new Museum recently erected near Wadham College. This collection, now called the "Bucklandian Museum," is particularly rich in cave bones. The history of caves in which the bones of extinct animals have been found was the Doctor's great forte, as has been well exemplified by his "Reliquiæ Diluvianæ," and other publications; and accordingly we find at Oxford,* cabinets full of the most gigantic shells

* I would call attention to "Cavern Researches, or Discoveries of Organic Remains, and of British and Roman Reliques, in the Caves of Kent's Hole, Ansti's Cove, Chudleigh and Berry Head," by the late Rev. J. McEnry, F.G.S., (Simkin, Marshall & Co., 1859,) in which are sixteen beautiful lithographic drawings, intended by Dr. Buckland to illustrate a second volume of "Reliquiæ Diluvianæ," which, however, he never published. By permission of Dr. Buckland's Executors, they are now given to

and bones of monstrous elephants, rhinoceros, bears, hyenas and wolves, which formerly roamed unmolested by man, not only over the greater part of Europe, but even over ground on which this great metropolis is built. Ages have passed away, but in their gravel charnel-houses they have left their bones; incontrovertible evidence of their once having inhaled the same air we now breathe, and pressed with their feet the same ground we now walk upon. These dry bones united into perfect forms by the acumen of Dr. Buckland, or touched by the magic wand of Professor Owen, again in these days start into life, grim ferocious monsters of the past: at our ease we study their forms and their anatomy; forms which, while evincing marvellous adaptation of means to ends, and Omniscient foresight, fill our minds with awe at the power of God, and the vastness of his creation.

But besides the collection at Oxford, Dr.

the public in aid of the Torquay Natural History Society, for the further explorations of Kent's Cavern, and Brixham Cavern.

Buckland also formed a private museum, which, at his death, shared the usual fate of private collections. It was transferred, in company with his scientific books, from the library and drawing room to the dismal, condemned cell of the auctioneer. Specimens that had been gathered by the same hand, from the same place, hundreds of miles away from home, and which had lain side by side in the same drawer many a long year, and which had been lectured on, disputed about, and admired by crowds of the most learned of *savans*, including, in many instances, even the great Cuvier himself, were now to be ruthlessly torn one from the other, destined never again to meet in their snug beds of cotton wool, and bedsteads of cardboard, canopied over by a gorgeous mansion of mahogany. In the auction-room I found the usual representatives of the various classes of society, deputations from the British Museum, and many learned Societies of the Metropolis; from the universities of Oxford and Cambridge; first-class dealers in minerals; and proprietors of stores in Wardour street, and swarms of smaller "dealers in curiosities," whose dens are to be found in out-of-the way corners, in out-of-the-way streets; and who came to buy anything that went cheap, either animal, vegetable or mineral, out of that large class termed the "miscellaneous." I found knowing old stagers in spectacles, who came to buy the pick of the auction, and who merely glanced at the trays as they were handed round; and green, young, ardent collectors, who seized hold of the trays as they were passed along, and retained them, examining specimen after specimen, to the annoyance of the tray-bearer, and the discomfort of others who also wished to examine. Lastly, I observed carpenters with paper caps and aprons, who came to buy the cabinets and the book-shelves, and not a few of the Hebrew race, without whose presence a London auction cannot possibly take place.

The first lots consisted principally of minerals from many parts of the world; of Sulphurs and

Celestines from Sicily, a very numerous collection, comprising beautiful and well-chosen specimens; most of them showing something different from their neighbors, and all of them in good condition—one specimen in particular was exceedingly beautiful; it was "a fine museum specimen of Sulphate of Strontian, or native sulphur from Sicily." It was a mass as large as, and about the shape of a lady's good-sized work box, and exhibited a natural miniature grotto extending three or four inches into the specimen, from the sides of which projected most beautiful needle-shaped forms of pure crystal, in some parts nearly meeting at the centre, and all glittering and resplendent.

The specimens of Fluors and Spars from Durham, Cumberland, and Derbyshire, attracted much notice; as well as some good examples of crystallized sandstone from Fontainebleau, which are now becoming more and more rare. Most of the minerals, I understood, were bought by Mr. Tennant, of the Strand, to complete the Stowe

Collection, which in the aggregate is the finest in England, and which we hear he intends to make quite complete, before he offers it for sale.*

After the minerals came several trays of Coprolites. Dr. Buckland was one of the first who made out the history of these curious formations. Along the sea shores at Lyme Regis, in Dorsetshire, peculiar looking stones were many years ago found in great numbers, and they were called "Bezoar stones," from their external resemblance to the concretions in the gall bladder of the Bezoar goat, once so celebrated in medicine; and they were imagined to be recent concretions of clay, such as are continually formed by the action of the waves, on clay-formed beaches.

* Mr. Tennant's Advertisement, 149, Strand:—"*Extensive and valuable Collection of Minerals.*—Mr. Tennant bought at the Stowe Sale the Duke of Buckingham's Collection of Minerals, which he has greatly enriched by a collection of Colored Diamonds, Australian Gold, and many other specimens of great value and interest. The collection, consisting of 3,000 specimens, is in two cabinets, each containing thirty drawers, with a glass case on the top for large specimens, and is offered at £2,000. Such a Collection is well adapted for any public Institution or any gentleman interested in mining pursuits."

Not so, however; Dr. Buckland proved that they were the fossil exuviæ of the extinct monsters, the Ichthysauri and Plesiosauri; whose fossil remains are found so abundantly at Lyme Regis and other places. In the Frontispiece will be seen the skeletons of these creatures gradually becoming covered with mud at the bottom of the water. Many of the specimens of coprolites at the auction were the very specimens used to illustrate the points insisted upon by Dr. Buckland, in his paper on this subject, published in the Geological Transactions, February, 1829, and also engraved in his Bridgewater Treatise.* Several of them were highly polished, and exhibited well the scales, teeth, and bones of the fish which the Saurians had eaten and half digested, some also contained the bones of small Ichthyosauri, which had been devoured by their cannibal relatives (see *Frontispiece*, figs. 1–5). In most of them was manifest the peculiar jet black color

* See New Edition. G. Routledge & Co., Farringdon Street, 1858, edited by myself.

which is probably the remains of the ink-bags of Sepiæ or Cuttle fish (see *Frontispiece*, fig. 7,) proving that the ever hungry Ichthyosaurus sometimes made his dinner off such small fry as cuttle-fish. Many, again, of these coprolites were unpolished, but not the less valuable for that; for they showed on their surfaces peculiar spiral marks impressed on them when in a soft state in the body of the devouring animal. Others exhibited, on careful examination, minute marks derived from the vessels of the intestines in which they were formed; thus demonstrating the structure of the mucous lining of the intestines of these præ-Adamite monsters. In order to follow this subject up, the Doctor had obtained and injected with plaster of Paris the intestines of dog-fish, skates, and other recent fish, which have the same peculiar spiral form of intestine as the fossil creatures must have had. Many of these recent specimens, prepared by his own hand, were placed in the tray side by side with the fossil remains, to act as interpreters of their real nature. Some of these coprolites have been turned to purposes of art, under the name of "Beetle-stones." Dr. Buckland had a table in his drawing-room that was made entirely with these coprolites, and which was often much admired by persons who had not the least idea of what they were looking at. I have seen in actual use ear-rings made of polished portions of coprolites (for they are as hard as marble); and while admiring the beauty of the wearer, have made out distinctly the scales and bones of the fish which once formed the dinner of a hideous lizard, but now hang pendulous from the ears of an unconscious *belle*, who had evidently never read or heard of such things as coprolites.

Dr. Buckland was, I believe, one of the first, if not the first, who called attention to the application of these coprolites for purposes of manure. In a paper published in the Journal of the Royal Agricultural Society, November, 1849, "On the causes of the general presence of Phosphates in the strata of the Earth and in all fertile soils;

with observations on Pseudo-Coprolites, and on the possibility of converting the contents of Sewers and Cesspools into manure," he writes: "Professor Liebig five or six years ago invited the attention of agriculturists to the possibility of applying to the same use as bone-dust, and guano, the fossil bones and coprolites which occur together in certain beds of the Lias formation. This invitation took place not many months after I had the honor of conducting him to the well-known bone-bed in the lower region of the Lias at the Aust Passage Cliffs, on the left bank of the Severn, near Bristol, where two beds of Lias (each from one to two feet thick) are densely loaded with dislocated bones, and teeth and scales of extinct reptiles and fishes, interspersed abundantly with coprolites derived from animals of many kinds which seem to have converted that region into the *cloaca maxima* of ancient Gloucestershire at the time of the commencement of the formation of the Lias."

Ever ready to apply his knowledge to practical

purposes, Dr. Buckland in the above paper, after analysing the probably chemical causes of the present condition of these coprolites, gives some valuable hints as the possibility of *imitating the natural processes that co-operated to their production;* and of converting the valuable phosphates of our sewers to the manufacture of a similar manure, by placing sewage water into conditions analogous to those which attended the formation of phosphates in the crag, and in other strata formed at the bottom of ancient seas and lakes.

Next in order in the list came a quantity of shells, both recent and fossil, from Hordwell Cliff, Lyme Regis, Isle of Wight, &c., and among them very good specimens of Ammonites (so called because they are coiled like the horn of Jupiter Ammon,) which had been cut and polished showing their internal structure; the animal itself had died and rotted away ages ago at the bottom of chaotic seas (*Frontispiece*, fig. 8.,) and the shell, remaining empty, became converted into a fossil. Several sections, &c. of recent

Nautilus shells were placed with them to illustrate the anatomy of the extinct shell by that of its still surviving relation. This tray brought to my mind the beautiful verses of the

AMMONITE AND NAUTILUS.

* * * * *

They sailed all day through creek and bay,
And traversed the ocean deep;
And at night they sank on a coral bed,
In its fairy bowers to sleep.

And the monsters vast of ages past
They beheld in their ocean caves;
They saw them ride in their power and pride,
And sink in their deep sea graves.

* * * * *

And they came at last to a sea long past,
But as they reached its shore,
The Almighty's breath spoke out in death,
And the Ammonite lived no more.

We next observed a tray full of bones of the flying lizard, the Pterodactyle (*Frontispiece*, fig. 4,) whose headquarters seem to have been situated at Stonesfield, near Oxford, for at this place their great long wing-bones are found pretty abundantly in the slate quarries.

It was often a puzzle to those who attended Dr. Buckland's lectures to understand how so many *separate* bones of these Pterodactyles are found. He illustrated this point by the dry body of a bird picked up on the sea-shore; as putrefaction advances, the bones become separated one from the other. The long wing-bones soon get loose, and, weighed down by the feathers attached to them (the pterodactyle had wings of a leather-like substance,) sink to the bottom, the rest of the body floating away.

Pterodactyle bones are also found in abundance at Lyme Regis. Among my father's MSS. I find the following letter* of his friend, and, I am proud

* MY DEAR BUCKLAND,

What think you of this for a motto?—

"Cesserunt nitidis habitandæ pisibus undæ,
Terra feras cepit, volucres agerabilis aer.
Sanctius his animal, mentisque capacius altæ
Deerat adhuc, et quod dominari in cætera posset."

Ovid, *Metam.* lib. i. fab. i.

If you can construe feras "crocodiles," and volucres "pterodactyles,"—and why not?—there is a pretty motto running on all fours as you shall see; winding up, moreover, with an elegant compliment to the Mayor of Lyme (if there be one) as the "sanctius animal," although he may resemble some of the *autochthenes* represented in your drawing (see *Frontispiece*) in his dining propensities. If you want to to say, my friend, the late lamented Mr. Broderip It is too clever to remain buried in a scrap-book. I trust the Mayor of Lyme will forgive Mr. Broderip's classical joke.

The great gem of the Stonesfield fossils, the jaw of the Phasclotherium, a small marsupial or pouched animal (hence such a big name for such a little creature *phaskolos*, a pouch, and *ther*, a beast,*) the first, and, at one time, the sole evidence of mammalian life having existed at the earlier period of the earth's history, when this slate was deposited, was not at the auction rooms; it has found a good, and, we trust, a lasting home

make it fit the Mayor exactly, you must adopt the reading of the Codex Meimobianus—

"Sanctius his animal ventrisque capacius alti
Deerat adhuc, et quod dominari in cætera posset."

Yours ever, my dear B.,
Raymond's Buildings. W. BRODERIP.

* There are many who are deterred from the study of Palæontology, or the science of things which existed long ago (*palai*, long ago,) by the hard names applied to the animals submitted to their notice. Let them only understand the meaning of the Greek names given to these animals, which is easily done by the help of Liddell and Scott's Dictionary, and they will find that these hard names are full of meaning, and the very best that could be used.

in the Museum at Oxford, but a few miles from the place where, ages and ages ago, it roamed over the neighborhood of Woodstock. Little did this tiny beast think that one day its under jaw would cause Dons to open their eyes with astonishment, and Professors to tax their memories and brains for appropriate words wherewith to descant upon its beauty, and upon the climate and state of animal and vegetable life at the time it existed.

Dr. Buckland had hammered and chiselled many of the above-mentioned fossils out of the rock with his own hand, and several of the hammers, &c., which he used for this purpose were included in the sale. There was a great competition for these among the buyers, and at last they were purchased, at a long price, by Mr. Tennant, of the Strand, for Mr. Sopwith, the engineer, one of my father's oldest friends. Mr. Sopwith afterwards very kindly gave two or three of them away to old pupils of my father's, who, although now occupying high places both in society and in the scientific world, have not forgotten the times when at Oxford they learned the A, B, C, of geology—literally "hammering" away at its rudiments in the stone-quarries of Shotover, Cumner, or Stonesfield.

These relics are much prized by the possessors, for by means of them my father hammered out much information from the breast of mother earth. They are, too, almost symbolical of the profession of the geologist. The Germans have caught up the idea, for I recollect, when a student at Giessen under Professor Liebig, that the young men who were studying mining wore two geological hammers stamped out of silver, in front of their working caps. This custom, I believe, is prevalent all over Germany.

After the hammers were sold there came a "dish of fish," but not such fish as we see displayed on the marble slab of Grove's—the fish now presented to our scientific appetites being *buried in*, not *placed upon*, slabs of stone. Whereas Billingsgate supplies, from the Doggerbank, and the submarine feeding grounds of sand plateaus deep under the waves of the British Channel, great cod, ling, turbot, and soles, and sends them to the cook in Belgrave Square but a few hours after they have been taken out of the water; our present bill of piscatorial fare, on the contrary, warrants the fish to have been brought from the Oolitic strata of Torre d'Orlonto, from the coal strata of Eislebon, from Lyme Regis, in Dorsetshire, from the Isle of Sheppy, and from Bohemia. Generally the fresher the fish the more is the money the vender obtains for the article; but, in this case, in proportion as the auctioneer could warrant the antiquity and staleness of his fish so did his hammer fall to longer prices. After the fish came the more solid parts of our intellectual repast. As a top dish was set before us a large box, full of gigantic bones of "the Mastodon, from South America," weighing at least half a ton; set off at the bottom by a dish composed of the bones of the Hippopotamus, from Sicily, teeth of ditto, garnished with portions of a fossil elephant's tusk.

For the side-dishes we had the vertebræ of a crocodile from New Jersey (*Frontispiece*, fig. 9); the bones of the ox, from Portland, and a fine jaw and teeth of the fossil Horse. For *entreés*, the bones and jaws of the fossil Kangaroo, from a cave in Wellington Valley, in Australia, of which every savant present seemed anxious to partake, for their great rarity caused the mouths of many a hungry collector to open and shut with emulous biddings, till at last they ran up to the sum of three pounds ten shillings, apparently a long price for a lot of such broken, meatless, and dirty-looking bones. Last of all, as pudding, we had passed round, plaster models, looking very white and pretty, and, moreover, much resembling the various shaped tarts and fancy cakes which the modern cook cuts out of pastry with sharp edged tin moulds. These were four boxes, containing "plaster models of microscopic living and fossil Cephalopods." Most carefully had they been prepared, and very pretty and inviting did they look; *plus* a little color, they would have passed muster as dummies of cheese-cakes and other palatable preparations in a pastrycook's window.

As dessert, were served up many trays full of preserved crab, not *apples*, but crab *creatures*, viz. crustacea; instead of candied ginger, beautiful white madrepores, whose appearance almost invited a bite; and for the paper crackers with elaborate patterns punched out in their ends, pretty lacework corals, whose patterns were far more delicate and far more beautiful, though produced by some of the smallest species of living atoms in the ocean, than the clumsy reticulations imagined by the brain of the biped creature that stands at the other end of the chain of created beings. Instead of pines from the hot-house, we had cones of the Araucaria excelsa, fossil vegetables from Chesterfield, with sections of silicified palm, showing the structure most beautifully. There were, also, several fine specimens of fossil Zamias, tropical plants not unlike a common pine-apple in appearance (see *Frontispiece*, fig. 10.) They had been procured by Dr. Buckland, in 1828, from the so-called "dirt-bed;" a stratum consisting of a dark brown substance, containing much earthy lignite, found immediately above the layers of Portland stone: the workmen know them by the name of birds' nests, their external form bearing a rude resemblance to the shape and size of a common crow's nest. It is a curious speculation, to think that these very zamias might have grown on the mud, which has since been converted by the hand of time into stone, and then, by the hand of man, into St. Paul's Cathedral. When I saw the aerial hut erected for the Ordnance survey, upon the top of the dome of this magnificent edifice, I thought to myself, that was not the first time since their creation, that these very blocks had supported a crow's nest."*

Among the "Miscellanea" was a very remarkable "brick from Babylon, with inscription and an impression of the foot of a dog." It ap-

* For full description of these Zamias see "Bridgewatei Treatise," vol. i. p. 453; and in various Plates, vol. ii.

pears that it was the custom of the brick-makers of old, to make their bricks in square moulds, and not in oblong shapes, as is done in modern brick-fields. These bricks were composed of mud, and placed out in the field to be dried by the *sun*, which, in eastern regions, is hot enough to do the work of a brick-kiln. That brick-making in those days was a most laborious and fatiguing task, is evident from the first chapter of Exodus, where one of the principal burdens placed upon the children of Israel by their taskmasters is thus mentioned: "And they made their lives bitter with hard bondage in mortar, and in brick, and in all manner of service in the field." In those days the reigning Kings did not put a broad arrow on government property, as is done now-a-days, but instead, they impressed, probably with an apparatus like a seal, a long inscription on the centre of the brick; in this instance, the inscription occupied about four inches by one. Whatever meaning the letters joining the words might have been intended to signify it is now impossible to ascertain; as, through time and the decay of brick, the characters are quite illegible.* When put out in a soft state, in order to get dry, it must have been placed on the ground; and when there, some careless and vagabond Babylonian dog, had placed his foot right in the centre of the inscription, and "his signature," written with his foot, remains to this day as perfect as the sign manual of the great Babylonian king, or even more so; for the marks of the two front claws and the double ball of the disloyal dog's foot are seen *obliterating* the letters of the regal inscription as plainly as if they had been done yesterday; it was not a very large dog either, for a few inches further back, and close to the edge of the

* I have lately been shown by Mr. Harle a pencil rubbing of the seal of King Darius (but which king has not been made certain.) The original is in the British Museum. The king is represented as hunting lions from a chariot. One lion is rearing up towards the king, who has an arrow fixed in a bow ready to shoot, while the coachman, or charioteer, is holding the reins. Another lion is seen defunct under the chariot-wheels. One cannot help associating this group of the figures of the King Darius and the lions engraved on a signet with the story in the Book of Daniel, chap. v., particularly when we read, "And a stone was brought and laid upon the mouth of the (lion's) den, and the king sealed it with his own signet."

brick, are seen the impressions of the claws of his hind foot, showing that his stride was about that of an ordinary sized terrier dog. I have compared it with the foot of a common black and white English fox terrier, and without knowing the history of the brick, it could be easily supposed that the footmark on it had been done last week by our trusty friend Pincher, instead of by a dog six hundred years before Christ.

The brick, having been dried in the sun by the Babylonian brick-maker, must have been taken up, dog's foot and all, and built, among other bricks, on the *top layer* of a wall, for at the bottom and on the sides, but not on the upper surface, we find, still adhering, a layer of bituminous asphalt; and this, when burnt, smells exactly like the bitumen we see sometimes being laid down in the London streets.

Now this asphalt was employed as cement to hold the bricks together; and, as our own bricklayers use cow-hair among their mortar, to make it bind more firmly together, so did the Babylonian workmen place between their layers of brick, and among the asphalt, reeds and straw. Again we turn to Exodus, where we find that straw was necessary to the children of Israel during their persecution by Pharaoh. "And the taskmasters of the people went out, and their officers, and they spake to the people, saying, Thus saith Pharaoh, 'I will not give you straw. Go ye, get ye straw where ye can find it: yet not aught of your work shall be diminished.' So the people were scattered abroad throughout all the land of Egypt to gather stubble instead of straw."

We examine the bitumen on our Babylonian brick, and we find beautiful impressions still remaining of the reeds or straw that had been placed in it for the purpose above-mentioned. These straws have made indentations in several places in the bitumen when it was put on soft, and probably hot; and in one of these, which is the size and shape of a slate-pencil, we can perceive even the cast of the parallel groovings in the siliceous covering of the reed.

When lecturing on the footsteps of the Cheirotherium, and other animals, Dr. Buckland always exhibited this Babylonian brick; and it was his wont, when commenting on it, to surmise that the inscription might be that of King Nebuchadnezzar, and that the dog who had put his foot on it might have been the property of the king aforesaid.

It was purchased at the sale, for the University of Oxford, and it is doubtless destined to be deposited in the new Museum there. We trust, that should this be read by any one who has the opportunity, they will examine for themselves this foot-print of "King Nebuchadnezzar's dog." Having learned that Colonel Rawlinson had discovered portrait models of the dogs used for hunting by the inhabitants of ancient Ninevah, I repaired to the British Museum, where my eyes were gladdened by the sight of these very dogs. Has the reader ever seen in the windows of the London shops where they sell sugarplums, almond toffee, and other variously shaped preparations of "sweet stuff," images of cows, horses, pigs, and dogs, made in red or white sugar? if so, he has seen almost the counterpart of these Nineveh models. These British Museum precious little relics, made of clay, are carefully locked up in a glass case, and there can be seen the five little models, as well preserved as though they were unbaked specimens of the genus dog, such as the gipsies use to put on the top of sticks at fairs and races, and at which we are invited to try our aim with thick sticks, at a small expenditure of the copper currency.

The dogs made of sugar, the china dogs from our English fairs and races, and the dogs from Nineveh, have all this one point in common, viz., that they do not stand upon their four legs, like the wooden horses on wheels that children ride upon, but upon a pedestal which fills up the space between the lower part of their bodies and legs: effect is sacrificed to trouble, for it is very difficult, as the writer well knows, to model an animal to stand on four legs without a support for his abdomen. The centre of gravity is sure to come in the wrong place. This fact became, from experience, apparent to the Nineveh artists as well as to the modern sugar-plum makers.

The animals of which these British Museum models are the portraits are, however, of greater antiquity than even the dog of the great King Nebuchadnezzar, for they lived in the days of Assur-bani-pal, B. C. 640. They were found at Nineveh, let into a slab, the sculptured figures on which represent a hunting scene, at which the above king is figured as being present. Their hair seems to have been cut poodle-fashion, or else they have a heavy collar round their neck. I could not quite make out for what the enlargement between the ears and shoulders was intended; it is more probable that a spiked collar was intended, such as are used at the present day by those lucky sportsmen who have the chance of hunting bears and other wild animals in their native fastnesses. Their shape, particularly about the fore-arm and body, is somewhat like the great, heavy half-pointer, half-cur dog used in Germany, and called "*jacht Hund;*" they all, however, seem to have curled tails, like the pugs, and heads more like the bull-dogs of the present day than any other of the canine race I can call to mind. They are all represented standing in an erect posture, with the eyes directed fiercely forwards, and realize the idea of a gigantic house-dog, who, having got to the furthest end of his chain, is savagely straining to attack an intruder.

Down their sides are impressed arrow-headed characters: these may or may not be the names of the dogs or their owners. I was not learned enough to read them.

In the same box with the ancient brick at the auction were two modern examples, from brick-yards in Oxfordshire; one showing the mark of a pig's foot, the other that of a cat, where these animals had been walking along. I have not unfrequently seen built into the walls of London houses, bricks marked at their ends with four thimble-shaped impressions, the moulds of the fingers of some idle boy, who, wandering about the brick-field when the bricks were yet soft, has left his mark to be perpetuated as long as the brick itself will last.

In the British Museum too, I found another parallel case with the foot-mark of Nebuchadnezzar's dog, but in this instance it was a cat, and not a dog, that had "put her foot in it." The cat had evidently been walking more circumspectly than the dog, for she has not put her foot on the royal stamp, but just by the side, or rather, in front of it, considering the direction she was walking in. This brick, upon which are both the regal mark and the cat's foot, is an ancient Egyptian brick, also made of mud and straw; it was probably put out in the sun to dry, and has handed down to us (were it otherwise wanting) the fact that there were cats in those days.* This brick, too, most likely was

* The great engineer Brunel, but a few days before his death, kindly sent me a mummy cat he brought from Egypt. I have taken the bandages off it, and find that it must have been a red-colored cat. Its dried-up tongue projects from its mouth after the fashion of dead cats of our own period.

placed on the top of other bricks to dry; for cats are fond of walking on the tops of walls, and probably paraded on the top row of the bricks of Thothmes or Rameses, as unconcernedly as they do now on the top of the brick stacks of Farmer Jones. I am also informed that there is a brick in existence which has on it the footprint of a Jerboa, one of the rat-like little animals common in Egypt even at the present day.

Among the bricks of the pillars under the Hypocaust (at the remains of a Roman Villa discovered at Wheatley, five miles N. E. of Oxford, 1845,) one was found with the marks of a boy's finger, another with the impression of a sheep's foot; presenting a curious illustration of the impressions of reptiles' feet on the new red sandstone. The tiles from the flues presented on their outside, many curious scratches of a dentated tool: these scratches varied with the caprice of the maker, and their use was to insure the adhesion of the mortar.

In an old basket in the auction-room were a

number of thin cakes of mud, simple, common mud, taken from a dried pond in the neighborhood of Reading. They looked valueless, but nevertheless were of some interest; for upon one side they exhibited numerous little round pits like the impressions of peas: these little marks had been made by drops of rain; the water had partially dried up in the pond, leaving the mud at the edges, soft, and easily impressed upon. A hard shower must have come on, for the rain-pits were pretty deep, showing that the drops had fallen with a considerable degree of force. Enticed by the rain, the earth-worms had come out, and had dragged their slimy bodies through the mud, leaving a long trail behind them. From these two circumstances we may conclude that all this happened in very hot weather; for why should the pond partially dry up, except from the evaporation of the water, leaving the edges almost dry? What could make the pit marks be so deep, except the heavy rain of a thunderstorm? A slight rain would have made hardly any marks. Then again, the worm came out on the wet mud for moisture. Lastly, the mud was contracted into saucer-like forms, and the edges of the impressions on it were very sharp, proving that after the thunder-storm, very hot weather came on again, causing the mud to contract and curl up in the above-named form. I have elsewhere seen the marks of a mole's comparatively broad, heavily nailed fore feet, and little delicate hind feet, in the half-dried mud in a drain in a ploughed field. He had probably come out to look for worms who were luxuriating therein. The markings of duck's feet, too, I have seen in the dried mud of a horse-pond, near Islip, in Oxfordshire, and have brought home a cake of it peeled off by the heat of the sun, with a duck's foot-mark beautifully impressed on it.

At Ruislip Reservoir, near Uxbridge, where I have been kindly allowed by Colonel M. Martyn to fish, there are extensive banks of mud left when the water is drawn off by the Canal Company; and on these are beautifully seen the footmarks of the geese, ducks, and herons that come down to feed there. It is most interesting to remark traces of the long, stately stride of Master Heron, as if he had been stalking along at his ease, and the spot where every now and then he had halted, faced the water, and made a lunge with his long bill at a fish. I can also trace where the little snipes have been running up the ditches; and can easily see the holes they have poked in the mud with their bills in search of food. The water-rats also have made numerous runs on the mud, and by the side of their holes, the ground is beaten down quite firm by their tiny feet.

In Connecticut, the fossil foot-prints of birds are found, and a detailed account of them has been published by Professor Hitchcock. There is an engraving of some of these in the "Bridgewater Treatise." In the Proceedings of the Ashmolean Society, at Oxford, an outline is given of their story:—At a meeting of the Society, November 14, 1842, Dr. Buckland communicated the results of some very curious observations by the late Hugh Miller, on the bottom of a large artificial lake, nearly sixty feet deep, at the base of the Pentland Hills, S.W. of Edinburgh, called the "Compensation Pond," which had been left dry by the droughts of last autumn. One of the phenomena in this pond, was the occurrence of recent footsteps in sand and mud since the pond had subsided. My father in giving the details of these footsteps, in his "Bridgewater Treatise," has supposed the cause of their preservation to have been, the passage of the animals over sand covered by a thin film of tenacious mud, which was punched through by their feet, as they crossed in the intervals between high and low water, over a broad expansive strand on the shores of the then existing seas.

When Professor Owen was lecturing at the Jermyn Street Museum, on the fossil-footsteps of birds, I made in a pie-crust and in soft clay several specimens to illustrate his lecture. By impressing the feet of herons, snipes, woodcocks, sand-pipers, &c., accurate footprints were formed, showing even the divisions of the skin on the under side of the bird's foot; if casts had been taken of these footprints, we should have had exactly the same appearance as the fossil "Ornithicnites" of America. There happened that day to be a dead ostrich at the College of Surgeons, which was under process of dissection; and with Professor Quekett's permission I made in plaster of Paris a beautiful impression of its foot. This is now in the British Museum.

In the Zoological Gardens, at Bristol, I observed a pseudo-fossil impression of a duck's foot, which I am convinced, would have taken my father's fancy immensely had he seen it. By the side of the duck-pond there was a margin of hard Roman cement. I suppose the ducks had not been driven out of the pond when the cement was put down, for one of them certainly had waddled out of the pond on to the land, and in so doing had put his foot in the Roman cement, which was then quite soft. It had subsequently hardened, and to this day this duck's foot remains impressed as firmly, and nearly as durable, as if it were a genuine fossil. I really think this ought not to be allowed to remain in its present position, but should be carefully cut out and put in a museum, to illustrate the science of footsteps.

The mud which, when moistened by the rain, is soft enough to take the common impression of a bird's foot, will, as can be easily conceived, give way to a considerable extent when a great heavy bull or horse puts his foot on it. Accordingly, on the bleak and marshy plain of Ottomoor, near Islip, we find numerous deeply-indented footmarks of cows and horses; these prove fatal to the young ducks; when following their mothers to and fro, in search of food, they tumble into these natural pitfalls, their soft bones and webbed feet not allowing them to climb, the poor little creatures cannot escape out of the hole, and there they remain till rescued by the farmer's wife, who is, from painful experience, well aware of what often becomes of her ducklings, and knows where to look for them; or else they quack away till night-fall, when a fox, weasel, or hedgehog, from the neighboring preserve, makes a meal of them and puts them out of their misery.

I have observed, in the Zoological Gardens, that after wet weather the elephant leaves enormous footprints in the mud round his swimming-bath; should a small child be unfortunate enough to fall into one of these, and be left to itself, it would be as helpless as the little duck in the cow's footprint.

These apparent trifles of every-day occurrence become important evidence when applied to the elucidation of geological questions. What happened at the Reading pond, happened centuries ago at the time that the old red sandstone was deposited, when the world was yet in its infancy. In the north gallery of the British Museum may be seen, against the wall, a very fine specimen of a slab from Storton Hill quarries, near Liverpool. On this are seen the footmarks of an enormous creature which has been pacing along the mud that formed probably the side of some antediluvian lake or estuary; just as at this day the mole crawls along a drain, and the duck paddles along a muddy bank, all distinctly leaving their footmarks. The mud upon which the ancient beast had trod must have been pretty soft, for his feet have sunk deeply into it, leaving the outline of the toes well marked. In one of the footprints, we can almost make out appearances that would lead us to conclude that the owner of the foot had found his progress rather slippery, for his foot seems to have turned slightly after he had placed it down. Such appearances may be remarked in our own footsteps on the London pavement, when the flagstones are greasy, as it is called. We may also remark in the slab, marks where cracks have taken place, just as cracks are visible in the mud of the dried pond.

Now, when these marks were first found, it was uncertain what manner of beast had made them; no bones were discovered that could be put down as having formed part of his skeleton.

What name, then, should be given to the creature? Its form, shape, and structure were quite unknown; but one thing was certain, that he had a foot in the shape of a human hand. Taking this, therefore, the only point certain about the animal, they christened it the Cheirotherium, or beast with a hand. Apropos to names given to fossil creatures, we cannot help here recording the fact that, after a long dispute among certain geologists as to what name should be given to an animal recently discovered, one wishing to give it this name, another that name, a certain learned and witty person proposed, that as it had caused a great bother in learned circles, it should be called the "Botheratio-therium." Again, when the reader next sees a picture of the extinct animal with tusks, somewhat like a walrus (only that they grow from the lower, and not the upper jaw), commonly called the Dinotherium, or dreadful beast; let him remark that Mr. Dinotherium is always represented as quietly lying down, like a cow chewing her cud, and always with his side towards the spectator. Why is this? The geologists have not yet ascertained after what fashion the hind limbs were formed, they therefore never represent the animal standing up, but always recumbent, leaving the form of the hinder limbs entirely to the imagination of the spectator, who, unless he be of a very inquiring mind, generally falls readily into the scientific trap.

As regards the marks made by the worms in the soft mud, we find from the "Bridgewater Treatise" that Dr. Buckland has compared these modern phenomena with those which took place during the Eocene period; and draws from these, in themselves, insignificant facts, logical conclusions as to the state of the atmosphere and water, in the very remote period of this planet when the Stonesfield slate was deposited. He writes: "We find on the surface of the slabs, both of the calcareous grit and Stonesfield state, near Oxford, and on sandstones of the Wealden formation in Sussex and Dorsetshire, perfectly preserved and petrified castings of marine worms, at the upper extremity of holes bored by them in the sand, while it was yet soft at the bottom of the water; and within the sandstones, traces of tubular holes in which the worms resided. The preservation of these tubes and castings shows the very quiet condition of the bottom, and the gentle action of the water, which brought the materials that covered them over without disturbing them."

I fancy I have discovered, as a pendant to "Cheirotherium," above mentioned, a remarkable creature which exists in our own time, and its habitat is somewhere in London. During last winter, I frequently observed in the London streets, when the pavement was covered with snow, or with mud that would easily take impressions, foot-marks of a very remarkable and puzzling kind. A creature wearing a human hob-nailed boot evidently passed along, but the creature had not left the "spoor," or track of an ordinary man, for the iron-shod heel was separated some two feet from the equally well armed toe, and the interspace between them was not marked with any impression, so that the animal could not have had a sole to his foot. In its journeyings, strange to say, it had *invariably walked backwards;* the turning up of the mud at the heel showed that, moreover, it had never walked in the center of the pavement or in the street, but always at the road-side in the pavement. For days and days I hunted this creature by its track, up Regent Street (it never went over the crossing,) down Oxford Street, both east and west, but I never could come up with it. Its home seemed, from the track, to be somewhere in the densely populated neighborhood of St. Giles'. As the Cheirotherium of ancient geological days was so christened on account of the track he made, I did not see why this Præ-Adamite animal, "the beast with a hand," should not have a modern cousin, "the beast with a foot," the true nature of both being uncertain at their first discovery. I therefore christened my mysterious creature "Podotherium," or "the beast with a foot," a provisional name, until I could make out its real name and nature. The snow melted and the mud was nearly dried up, and I thought I should never solve the mystery as to what Podotherium was. At last, one day, after a heavy shower, I crossed the tracks very strongly and recently marked, by Wellington Street, Strand; I followed them instantly, and found that this mysterious creature was only a poor old cripple, who, having lost his feet, could not walk upright after the ordinary fashion of men; he could only crawl along upon his knees, after the fashion of the monks of old, who have worn away with their ascending knees the stone steps of the shrine of Thomas à Becket in Canterbury Cathedral. Necessity, in Podotherium's case, had been the mother of invention. He had cut an ordinary hob-nailed boot in two, had affixed the heel part on to his knee, and the toe part on to the place where the foot ought to have been: so that as he crawled along upon his knees, the heel part of course had been going foremost, giving the appearance of a man walking backwards. "Podotherium" was gaining his livelihood by selling to passers-by tin funnels, which he carried in a string around his neck, and he therefore always kept, as his footmarks indicated, to the roadside of the pavement to solicit custom. As Cheirotherium in time was proved to be a gigantic sort of a frog (whose restoration we may see, by my friend, Mr. Waterhouse Hawkins, at Sydenham Crystal Palace,) and now is known by the name of Labyrinthodon, on account of the labyrinth-like structure of the teeth; so "Podotherium" proved to be simply poor old —— of Seven Dials, the tin funnel-maker. This story may seem absurd to many of my readers; but I would beg them to recollect what consternation was caused by the newspaper accounts of some mysterious footmarks in the snow in Devonshire, not very long ago.

It also shows that observation of common things is necessary to the elucidation of unusual natural appearances,—and that, by being awake to the commonest things that take place around us, we are enabled to interpret unusual phenomena when they arise, whether the case be the fossil track of an animal which has long ceased to exist, the track of a burglar who has broken open the clergyman's house and taken all his plate, or the clearing up of a murderer. Take, for instance, that most remarkable instance of the "Waterloo Bridge murder," where a number of human bones were found cut up, semi-pickled and boiled, put up in a bag, and all placed on a jetty of the bridge. By the kindness of the authorities I was enabled to inspect these bones, more than once, in company with Professor Quekett. The only things that could be predicated from previous observation were that the cuts had been made by a man accustomed to handle a saw; that it was not done by a person who had a knowledge of anatomy—the merest tyro with the *scalpel* would have gone to work differently; and that somehow a woman and a cat were mixed up with the affair—for Professor Quekett and myself found the long hairs of the former, and the short hairs of the latter, sticking on to the semi-pickled and afterwards boiled bones. There were also interesting conclusions to be drawn from the appearance of the clothes, &c., whereby it was made quite clear *how* the body was dissected; but it required quite a different series of reasonings to ascertain *who* committed the murder.

Mr. Galton, in his most interesting book, "Tropical South Africa," Murray, 1853, records a barbarous semi-murder, which would never have been discovered but for some very remarkable tracks on the soft sand of the desert, which immediately struck his observant eye. He writes:—"I saw a horrible sight on the way, which has often haunted me since. We had taken a short cut, and were a day and a half from our waggons, when I observed some smoke in front, and rode to see what it was. An immense blackthorn-tree was smouldering, and from the quantity of ashes about, there was all the appearance of its having burnt for a long time; by it were tracks we could make nothing of; *no foot-marks, only an impression of a hand here and there.* We followed them, and found a wretched woman most horribly emaciated; both her feet were burnt off, and the wounds were open and unhealed. Her account was, that many days back she and others were encamping there; and that when she was asleep a dry but standing tree, which they had set fire to, fell down and entangled her among its branches: there she was burnt before she could extricate herself, and her people left her. She had since lived on gum alone, of which there were vast quantities about; it oozes down from the trees and forms large cakes on the sand. There was water close by, for she was on the edge of a river bed. I did not know what to do with her; I had no means of conveying her any where, or any place to convey her to. The Damaras kill useless and worn-out people, even sons smother their fathers; and death was evidently not far from her. I had three sheep with me, so I off-packed and killed one. She seemed ravenous, and though I had purposely off-packed some two hundred yards from her, the poor wretch kept crawling and dragging herself up to me, and would not be withheld, for fear I should forget to give her the food I promised, &c. I did the only thing I could. I cut the meat in strips and hung it within her reach, and where the sun would jerk (*i. e.* dry and preserve) it. It was many days' provision for her. I saw she had water, firewood, and gum in abundance, and then I left her to her fate." I myself have more than once remarked the track of a man with a wooden leg—it has a most curious appearance—in the snow or mud; the mark of the human foot catches the eye directly on the one side, the round hole made by the wooden leg on the other. There was an old verger at Westminster Abbey, in Dean Ireland's time, who hated any person with a wooden leg, and no persuasion would allow him to let the one-legged visitor into Edward the Confessor's Chapel. The verger had a reason for refusal: the floor of the chapel shows remnants of a beautiful tesselated pavement, and the old verger said that wooden-legged people "*punched* out the few remaining bits of the pavement with their wooden *stumps*." He would therefore never let them in.

As if in explanation of the science of natural casting, there were put up for sale at the auction some very pretty natural casts from the baths of San Filippo, between Rome and Sienna. The history of these casts is as follows. The water at the above named place is very highly charged with carbonate of lime; and the fact of its depositing the mineral mater it contains in solution, upon sticks and grass, thus taking accurate casts of them, having been observed, certain ingenious persons have placed under the drip, moulds of medallions of heads, antique figures, &c., made of sulphur; the water, careless of results, artistic or not, has deposited carbonate of lime in the mould to the thickness of half an inch or more, taking a most beautifully accurate cast of the figure in relievo, the surface being very smooth and polished, answering to the surface of the sulphur. This deposit goes on so gradually, and with such minuteness, that even the lines in a fine wood engraving have been accurately moulded, and we have the picture in hard solid carbonate of lime, instead of thin perishable paper. If we reverse one of these stone pictures, we shall find that the outside layer is exceedingly rough and indented, the results of the water dripping from the well. The process of the formation of fur in the tea-kettle, is only another instance of this process, except that the deposit is not applied to useful and artistic purposes. The best way, by the bye, to keep off this fur, is to place a common marble in the kettle, which by perpetually rolling about prevents the fur forming; or else to "put in a clean oyster shell, which will always keep it in good order by attracting the particles of earth or stone:" but I have never tried this. In the neighborhood of Oxford this deposition of fur is not uncommon. There is in the Museum there about a foot of lead piping, which was taken up because the water would not flow properly through it. The cause of the obstruction was soon ascertained: a horse-chesnut, with its rough spiny coat, had got in; the spines became jammed between the sides of the pipe, and there it had remained, forming a nucleus for the deposition of the carbonate of lime, which had formed quickly and densely, completely obstructing the pipe. In the same Museum is suspended against the wall, what at first sight appears to be a wooden pipe, but upon examination it will be seen to be not a wooden but a stone one; it is a stone pipe formed within a wooden one. The original pipe, about four inches square, belonged to the old conduit at Carfax, in Oxford; in process of time the carbonate of lime had formed equally on all four sides of the pipe to the thickness of about a quarter of an inch. When taken up, the wood-work was found to be quite decayed,* but, nevertheless, it did not leak, as

* I learn that the fashion of pollarding, or cutting off the

there was another pipe within it, firmer, more solid, and more lasting than the wooden one, through which the water flowed; this second pipe is of stone, naturally formed within the first.

On examining our stone specimen we find that the inside is smooth, but the outside, on the contrary, is quite rough, preserving accurately the lines of the rough fibre of the oak from which the wooden pipe was made, and in one place a mark, very like a saw-mark can be traced.

In the British Museum, North Gallery, No. III., may be seen just such another naturally formed stone pipe. It was obtained nearly one hundred years ago, out of a pipe that conducts the waste water from a coal pit at High Little-

branches of elm-trees in the neighborhood to make them grow tall and straight, arose from the former demand for their stems, cut into lengths, and bored throughout, to make pipes to conduct water. I have frequently seen these in the London streets when the workmen have been making excavations for repairs. Coffins are generally made of elm, because they last longer in *damp* places than any other wood. Oak coffins last longest in dry places, such as family vaults above ground, &c.—*Auct. an Undertaker.*

ton, between Bristol and Wells, and a description of it will be found in the "Philosophical Transactions," 1773. The deposition is in laminæ of various colors, some light colored, some dark colored. In one place a nail has projected from the outside to the inside of the board, over this the lines of deposition have taken a curve; on the outside of the pipe may be seen roughnesses, and marks of the knots in the elm board of which the wooden pipe was formed. This pipe was fixed in 1766, and was obliged to be taken up in 1769.

The describer of the above states that in the churches of Lima, South America, are many statues and beautiful holy water basins, formed entirely by deposition, moulds being placed in a petrifying spring in the neighborhood. This is indeed going a step further in applying natural formations to art, than the good people at San Filippo have yet done; statues and fonts could probably be made as well there as in South America.

As regards the difference in color in the deposition of the pipe, one authority states that it is caused by ochre, but then why was there not always ochre in the water? This phenomenon may most probably be explained by the fact that the black marks were made on the days when the men were working in the mine, and by their operations made the water dirty, which of course then deposited a black mark; on the days when no work was doing in the mine, the water would run out clean and deposit a white mark.

In the Oxford Museum is a fragment of calcareous deposit from a pipe, formerly attached to a coal mine in the north of England; the marks upon it tally exactly with the manuscript ledger of the overseer. We find black and white marks alternately, six of each, i. e. the men worked by day and rested by night: then we have a mark the size of a black and a white one put together, which mark is quite white; it was Sunday, and the men rested twenty-four hours: so goes on week after week, black and white marks alternately, a double-sized white one for Sunday. In the middle of one week we find a Sunday mark; how came it there? The reason is this—the races were held at a neighboring place, the colliers all left off and went to the races, and did no work; but they made up apparently for lost time afterwards, for our stone indicates by a subsequent double black mark that they worked day and night. This is properly called "the Sunday stone," and its history was a favorite story of the late Dean's.

Another illustration of deposition was also put up for sale at the auction; it was "a large and perfect amphora, encrusted with shells from the bed of the Tiber." This amphora was of great age, and had evidently reposed for many centuries in the mud at the bottom of that river, for on it were seen in clusters, in different parts, from eight to ten common oyster shells, which once of course contained living inhabitants. The amphora itself was of the usual elegant shape, with a handle at each side for lifting it, and the bottom tapered off to a sharp point like the conical hats of ancient times. The Romans, who kept their wine in these amphoras, were accustomed to place them in sand, which necessitated this pointed bottom, and the handles were admirably placed, to give a purchase to the person who screwed it into the sand. The workers in hardware of the present day, I see, have imitated this form for holding flowers, &c., but they have been obliged to fasten three of them together, tripod fashion, in order to make them stand upright. Now, it is more than probable that this amphora had taken its upright and proper position in the mud of the Tiber, for the oyster shells were fixed not towards the lower part, but all on the uppermost part; they could not get to the parts below the mud, but readily fixed themselves to those above it. Dr. Buckland found this amphora in a corner of an old curiosity shop in Rome; purchased it, and for many a long year it decorated his hall both at Christ Church and at Westminster.

In the British Museum, in the North Gallery, No. III. Case 46, may be seen a still more remarkable object from the bed of the Tiber. It is a human skull, which has become incrusted all over with a coating of carbonate of lime. Not only is the outside completely covered, but the inside of the skull is also concealed by it; the knobs and eminences therein existing, having received their share of the deposit, stand out like little mountains amidst valleys. Could we take off a portion of the inside lining, we should most probably—nay, even we might say, most certainly—find accurate castings of the grooves which are hollowed out in the bone, to contain arteries and veins. The features of the skull are hardly distinguishable; and from the uneven forms of the deposition about the bones of the face, it looks like a skull with the mumps.

In the same case may be seen other specimens of incrustations, such as a bird's nest and eggs from the wells at Knaresborough, and also a few examples of the San Filippo natural casts. In the same gallery, Case 21 contains some remarkably beautiful incrustations of pure white deposit upon a twig from the hot springs of Roturia, in New Zealand, which were presented by Sir George Grey. The twig looks as though it had been gathered from an English hedge on a bright frosty morning, and the hoar frost had suddenly become fixed permanently upon it, in defiance of the powers of warmth to melt it. In the same case is exhibited the cast of a feather-moss and other substances from Lazon, one of the Philippine Islands.

While walking lately in a garden near Windsor, my attention was attracted by the end of a stone projecting from some rock-work, made of flints, corals, biss of iron and brick, rounded quartz, pebbles, broken portions of ammonites, and other sundries, which were artfully piled together, and dignified by the name of rock-work. The stone I observed was none of these; the portion projecting was about three inches in diameter, rounded at the edges, and concave, like the vertebra of an ichthyosaurus, which I at first imagined it might be, particularly as it was covered with a thick coat of whitewash, and its structure concealed. I received permission to examine it, and accordingly ruthlessly removed the top stones of "the work," and obtained the specimen. Stone it certainly was, but a stone of no geological formation, as I could declare the moment it came into my hands, from its lightness, unless it might be a bit of pumice-stone; but then who ever saw a bit of pumice-stone in the shape of a three-sided pyramid, whose base was round and concave, as if water-worn? yet it could not be water-worn, as the three sides of the pyramid were separated by a well-marked ridge between them; the sides, too, being quite flat, as if they had been in opposition to another stone. All parts of the stone were covered thickly with a coat of whitewash, which quite concealed its structure. The stone was kindly presented to me. The pump soon took off the whitewash; and then, from its peculiar external structure, it became evident that it was a calculus, or stony secretion from the inside of an animal. To a human being it was much too large to have belonged, for it was as big as a good-sized cocoa-nut; it probably, therefore, had been taken from a horse. Down one of its sides there was a slight crack: having placed it on a table, I carefully inserted a chisel into the crack, and after a few gentle taps on the chisel with a hammer, an outer case fell off, about one quarter of an inch thick, leaving a second calculus inside, like a little box found inside a big box in a so-called nest of boxes. The structure of this second calculus was different from the external one; instead of being fragile and porous, it was composed of a dense and hard limestone-like material. With some little difficulty, and with great care, by the aid of a fine saw, I made a section. My trouble was rewarded; for, as we had predicted to a by-stander as likely to be the case, we found a nucleus, which nucleus was evidently a bit of lead. The section at the same time revealed layers symmetrically placed one upon the other—the first layer being formed round the bit of lead, the second round that, and so on, exactly as a schoolboy makes his great snow-ball the nucleus of his snow-man, by rolling it along the snow-covered ground; layer after layer is added, and if a section be made when the schoolboy is tired of his work, the snow-ball will be found to be formed by concentric layers, having for its nucleus, probably, a bit of snow hardened and well kneaded by the hands of the operator.

To a horse, therefore, I concluded my pseudo-fossil had belonged. I made inquiries in the neighborhood about its history, and learnt from one of the servants, that some years ago, the house belonged to the man who owned the mills at Clewer, about a quarter of a mile distant This miller had a horse, which worked about the mill, drawing the sacks of corn, &c. The horse died suddenly, and the miller, anxious about his old favorite had him opened: inside him were found three stones, one of which had deen placed in the rock-work, by way of ornament; the others had been thrown away. Thus then, I account for the marks on my specimen, which had told me that it had been in juxta-position with something which was not very hard, and which fitted on to it. But, to account for the nucleus of lead. It is most probable that this lead was a crushed shot, which had got in among the horse's corn, and been taken up with the corn into his mouth; his great flat grinders had crushed it in their grip, and flattened it, and down it had gone along with the rest of the food: an insidious enemy, destined, in course of time, to cause much pain to, and ultimately the death of the miller's poor horse. In order to verify my notion, I crushed several shots with hammers and between heavy stones; the shots came out presenting very much the appearance of the leaden nucleus. From my observations, it is probable that the size of the shot was that commonly called partridge shot.

Almost the last lot in the auction catologue was "A stuffed Hyæna on stand, twenty-four years in the Surrey Zoological Gardens." In itself this signified not much, but there is a history connected with this animal. When Dr. Buckland was writing his essay on the celebrated bone cavern of Kirkdale, in Yorkshire, he took great pains to compare the bones there found with recent bones, so as to make his story quite complete. In the Kirkdale Cave he found a portion of skull which he believed belonged to a young hyæna, and although nearly certain that it was what he thought it to be, he ransacked all the collections he knew for a recent skull of a young animal for comparison; and not finding one, he requested Mr. Burchell, the great African traveller, to send him a young hyæna from the Cape. In course of time the baby-beast arrived in the docks; a pretty tame little beast, a great favorite with the sailors, who had christened him "Billy," doomed nevertheless to be slain for the sake of science. The late Mr. Cross, then of Exeter 'Change, and afterwards of the Surrey Zoological Gardens, acted as agent, and undertook the delivery of poor

"Billy." The little brute, however, by his good temper and playful manners, quite won the heart of Mr. Cross, who begged hard for his life, and at length obtained a remand on condition that the skull of a young hyæna should be forthcoming. Mr. Cross, we suppose, turned out all his drawers and cabinets in search; anyhow, he produced a skull within the given time, but it was not the skull of poor Billy. His life was spared, and he was forthwith taken to Exeter 'Change, and thence removed with the rest of the wild beasts to the Surrey Zoological Gardens. I have in my possession "The Companion to the Royal Menagerie, Exeter 'Change, containing concise Descriptions, scientific and interesting, of the curious foreign Animals now in that eminent Collection, derived from actual observation, by Edward Cross, Proprietor, 1820." The following is the description of poor Billy, then in his youth, but amiable withal:—"The hyæna in a cage at the end of the room is possessed of a large share of good humor, and entertains the visitors at feeding time by the gesticulations of delight he manifests at the moment, and by his curious imitations of the human voice resembling laughter. This animal suffers himself to be caressed, and is so familiar with the keepers, that when any repairs are wanting to his cage they have no hesitation in going in with him. (N. B. This was before the days of Van Hamborough, and other Lion Kings.) He is a native of the Cape of Good Hope, and is frequently called the Tiger Wolf." Billy arrived in England in the year 1821, and he died in his den a peacable and quiet death, January 14, 1846, having lived just a quarter of century within this metropolis. I remember well when quite a child being taken to feed Billy with cakes, and of later years I paid frequent visits of inquiry, always with satisfactory results; he was always in a good humor when we called on him, expressing his delight by spinning round and round in his den as fast as he could manage to go without falling down, for he was getting old and decrepid; all this time he uttered that remarkable sound peculiar to hyænas, which once heard can never be forgotten, and which sounds so much like a laugh proceeding from human lips.

At his decease (the cause of death, *plus* old age, being an enormous goitre in the throat,) Dr. Buckland presented his carcass to the Royal College of Surgeons, reserving, however, the skin for himself. Mr. Bartlett, the Taxidermist, undertook the stuffing of it, and very cleverly made a plaster cast of the head directly after the skin was removed, leaving the muscles, &c. still attached to the skull. This cast, when placed into the skin of the head, of course accurately fitted it, preserving to a certain extent even the expression of the animal. This, then, is the history of the "Stuffed Hyæna at the auction." The carcass was sent to the College, and was carefully dissected, many of the internal parts being preserved. Professor Quekett, who dissected it, informs me that his flesh was as hard and tough as a bit of shoe leather, and that he was obliged to put a fresh edge on his knife after every two or three cuts. After dissection the bones were made into a skeleton, where it may now be seen, No. 4,446, in the Osteological Gallery. One of the canine teeth is worn down quite to a stump, and the grinders are covered with tartar; one of them he has cracked in some of his bone-splitting operations, for the jaw around exhibits symptoms of former inflammation. On the head may be seen lines and furrows in which were inserted the great powerful muscles which moved the under jaw. His legs, crooked and bandy in life, preserve the same appearance in death. Poor Billy first made his *début* as the youngest hyæna in England; he ends his career grim and grisly as the oldest hyæna in England, and probably in Europe. The stuffed skin is now at the College of Surgeons in company with his skeleton, having been bought at the sale by Professor Quekett.

Not only was Billy subservient to the cause of science when dead, but even when alive he unknowingly gave much important assistance to his former owner, then busy with the "Reliquæ Diluvianæ," for Billy cracked the marrow-bones of oxen, and refused those bones which contained no marrow, exactly as did his ancestors ages before him in the wilds of Yorkshire, as yet untrodden by the foot of man. So wonderfully alike were these bones in their fracture, that, judging from this point alone, it was impossible to say which bone had been cracked by Billy, and which by the aboriginal hyæna of Kirkdale. Again, Billy polished with his feet and hide the sides and floor of his den of wood, as his ancestors did the sides and floor of their den of stalactite in the Yorkshire hills; and as the ancient beasts deposited "album græcum" in abundance after a dinner of bones, so did Billy deposit pounds of the same substance; even in this minute circumstance illustrating the history of his ancient British forefathers. The details are recorded in Dr. Buckland's "Reliquæ Diluvianæ."

After vast labor and much accurate observation, Dr. Buckland at length made the evidence of the former existence of hyænas in England quite complete; so complete indeed, that on one occasion when surrounded by the actual bones and specimens knocked out of the Kirkdale stalactite by his own hand, and sitting in his Professor's chair in his own museum, he appealed to one of the most learned judges in the land, who happened at be present at his lecture.

After having, with his usual forcible and telling eloquence, put his case, to prove not only the former existence of hyænas in England, but even that they were rapacious, ravenous, and murderous cannibals, he turned round to the learned lawyer and said, "And now, what do you think of that, my lord?" "Such facts," replied the judge, "brought as evidence against a *man* would be quite sufficient to convict and even hang him."

APPENDIX TO THE ABOVE ARTICLE.

Dr. Buckland was particularly careful to put descriptive labels on all specimens that came into his possession; the moment a fossil, a bone, or other specimen, was transferred from the road-side quarry, or from the workman's cottage, immediately was written on it in ink the name of the place whence it was procured, and also the date when it came to hand. This we find on many Saurian bones, teeth, vertebræ, &c.:—"Lyme Regis, 1840;" "Shottover, 1824;" "Val D'Arno, 1826." The task of labelling generally fell to the lot of my beloved mother; she was particularly neat with her fingers, and accurate with her eye, and had the knack of picking out good places on the specimens whereon to write the labels, without interfering with the appearance of the specimen itself (a no small art); she likewise, from long practice, so formed the actual letters that, in many cases, the words appear more as if printed than written with the hand. I cannot sufficiently urge upon all collectors, whether of surgical, pathological, antiquarian, geological, or zoological specimens, whether for use or as simple curiosities, the absolute *necessity* of labelling each individual article. It will not suffice, and it is in many cases next to useless, to put on a paper label; for consider what happens in a few years (as I well know to my cost)—the gum or other cement cracks, and off goes the label. The specimen is found without a history. Take, for example, a tooth of a fossil Hippopotamus; unlabelled, it is not much more than an old bone, but when its locality is written on it,—say, in this case, "Sicily," or "Charing Cross,"—for Hippopotami bones are found in both places,—and the specimen becomes immediately available as affording evidence which may be at some time and in some place exceedingly valuable. This plan may be well illustrated by what has happened within the last few weeks, when a great talk has been made about "Man among the mammoths." My brother recollected that there was a box taken down to my mother's house at St. Leonard's, containing "flints," &c., and he thought there might be in it some of the very ancient stone arrow-heads, which being found mixed with bones of the primæval elephants, mammoths, &c., have lately caused so considerable a sensation in the scientific world.

This box was sent up to me, and I was much pleased to find a beautiful series of fractured flints, collected by my father. Apropos to this story, and the conchoidal fracture of flints, from gun-flint manufactories, &c., I found also a large collection of stones used by the Druids to put a polish on the stones at their ancient temple at Carnac, together with a pretty little collection of stone hatchets, used by savages of the present day, all brought together (and Dr. Buckland must have had some trouble to do this), to illustrate what is certainly the nucleus of the whole box, a very fine flint arrow-head or axe-head, seven inches long by five wide, with this label in the Dean's MS., *written in ink upon the actual surface*, "Hoxne, Suffolk, found in loam with bones of elephant. See Archæologia, Anno 1800."

These few words give authenticity and value to what appears to be a simple bit of broken flint-stone, *per se* worthless, but *with its history* exceedingly interesting and valuable.

At the top of the box are a number of papers, old letters, scraps of notes, and pencil drawings, which form a key to the contents of the box: and this fact emboldens me to urge upon my readers the necessity of putting all papers relating to specimens, with *those* specimens, as well as writing on the specimens themselves, or, if this be not possible, paste them into a scrap-book, with numbers corresponding to the specimens they refer to. I must give an example.—In the Oxford

Museum is a boat-scoop, made of the tusk of a fossil elephant. It was purchased by Captain Beechy, from the Esquimaux, and my father has pasted *inside* it a most interesting letter relative to its history, giving it authenticity and great value. No one who has visited the Museum of the Royal College of Surgeons, Lincoln's Inn Fields, can fail to be struck with the beauty, rarity, and orderly arrangement of the numerous specimens placed within the glass cases and on the shelves. According to the last census, there were no less than 44,701 preparations; each specimen is labelled, and there is no difficulty in finding its history, and all that is known about it, in the pages of the catalogues. To get the Museum to this state of perfection has caused a vast amount of labor; but now, thanks to Professor Owen and Professor Quekett, admirable order has been produced out of what was comparative confusion some years back. Hear the words with which Professor Owen begins his preface to the volumes of the present catalogue:—"The surviving friends of John Hunter well remember how deeply the latter period of his life was embittered by reflections on the imperfect condition of the records and catalogues, so essential to the value and utility of his collection. It appears to have been too much a habit throughout the whole period of its formation, to trust the history of the specimens to memory; and the absence of any adequate system of notation or reference by which they could be recognised was severely felt when the powers of mind which had called them into existence began to be shaken, by the reiterated attacks of a severe and ultimately fatal disorder."

Luckily, however, sufficient evidence to the identity of the preparations remained, but it had to be gleaned from various sources with great labor. The present indefatigable and amiable Curator of the College, my good friend Professor Quekett, is daily lamenting the want of labels and descriptions upon different preparations which from time to time turn up. "Ah," said he, "if people would only write upon their specimens, or give references to them, I might have been saved hundreds of hours of hard work."

Professor Quekett showed me some little boxes, which he finds of the greatest use for small specimens, which otherwise would be folded up in a bit of paper, put away, and, in ordinary hands, probably get mislaid; they are very light and strong, made of the best card-board, and the tops are of glass, so that at a glance the contents of the box can be seen. I find Mr. Tennant, 149, Strand, has lately added a stock of these most useful articles to his shop, and I can confidently recommend them. They are called glass-capped boxes for eggs, shells, fossils, minerals, medals, objects of antiquity, jewelry, &c.; there are eight sizes of them, and Mr. Tennant has a circular, showing the various sizes, and the prices, which are very reasonable.

This practice of labelling may be applied to every-day life, to keys, knives, &c. You find a bunch of bright keys in the street; you know that somebody somewhere must or may be, at that moment, distractedly turning out all his pockets, and hunting his wife and servants all over the house to "look for my keys." We also see advertisements in the *Times*, nearly every morning, for lost keys. Now if there had been any notice affixed to the keys, showing who was the proprietor, all this anxiety might have been avoided. Now, it does not do to put the full name and address of the owner upon the bunch of keys, for Mr. Housebreaker might pick them up, and take the opportunity of using those keys for purposes exceeding the bounds of simple curiosity. He might open the street-door (the address on the keys would tell him where to go,) and pay a midnight visit to the master's writing-desk, cash-box, the cellaret, or may be the plate closet.

All the disagreeable chances may be avoided, and yet the keys recovered, by a simple plan, told me by my friend Mr. T. L. Coulson, of Bristol. It is this:—have a bone or ivory label affixed to the keys, with an inscription such as this, "Post-office, Charing-Cross, box 13, five shillings reward." Or if the owner does not live near Charing-Cross, let him put on the label the name of the nearest post-office, or tradesman's shops where he is known; still adhering to box 13 (or any number he chooses) and the reward offered; or he may have the name of his club, or the post-office of his country town inscribed, though he may not have a box at all there. For my own part I have on my keys, knife, &c., "Athenæum Club, Pall-mall, five shillings reward." Now, how does the plan work? The keys are dropped you know not where, but the chances are somebody finds them; he reads that he will get five shillings by taking them to the place indicated, and the keys will probably be taken there at once, as they are of no use except for old iron. In the meantime, having discovered your loss, you should go to the post-master, or the club-porter, say that "box 13" is on the label of a bunch of keys, which you trust will be brought to him in the course of the day; deposit the reward money, with a fee for the receiver; and the chances are your keys will be in his hands in twelve hours' time, and you will have saved yourself a monstrous deal of anxiety and fretting.

Visitors to museums are very apt to touch and handle specimens,—this is an itching which seems natural to us all. Dr. Buckland had on his drawing-room tables, at Oxford, many things very pretty to look at, and valuable in themselves. Through frequently handling, they had, from time to time, been injured and broken; he therefore placed on these tables labels with the words, "Paws off," conspicuously engraved on them. This concise mode of expressing the wish of the proprietor had the desired effect, and paws were kept off.

Dr. Buckland could not bear to lose an umbrella (they were never very good ones of their kind, but still he could not bear to "lose his umbrella.") He lost two or three in one way or another, and at last he had inscribed, in conspicuous letters, on the handle of a new one he bought purposely, "Stolen from Dr. Buckland," and this he never lost. It was fairly worn to a skeleton through long and faithful service, till at last it became so very shabby, that I often wished somebody *would* steal it; but the large label, "Stolen from Dr. Buckland," kept away everybody, nobody even ever offered to "borrow" it on the wettest of wet days, although it often invitingly stood by itself, in solitary glory, in the umbrella-stand in the hall.

Dr. Buckland, at the beginning of his lectures on Mineralogy, invariably taught his class how to handle minerals, &c. "Always take specimens carefully *by the edges* between the finger and thumb," said he, "and put them down gently directly you have finished examining them, and don't keep them in the hand when you are thinking about other things."

Having mentioned paste and labels,* I would here beg to give the receipt for the paste which my father always used, and which may be useful to some of my readers (if they did not know it before ;) it is copied from my mother's MS. :—

DR. BUCKLAND'S CEMENT.

1 part finely-powdered white sugar,
3 parts finely-powdered starch,
4 parts finely-powdered gum Arabic.

All by weight. In mixing use cold water. Rub the above ingredients (dry) well together in a marble mortar ; then, by very little at a time, add the water till it is the thickness of melted glue ; put it in a wide-mouthed bottle, and cork closely.

Many of the larger bones and fossils at Oxford have been mended with a whitish-colored cement. which is exceedingly hard and tenacious. My poor mother took great delight in mending and putting together broken fossils which the Dean

* Good labels can be bought at 18 Holborn Hill, E. C.

considered valuable enough to have the trouble spent upon them. The quarrymen were often very careless in getting the specimens out of their matrix, and would bring a good fossil much in the state of a London street flint when a great coal-wagon has passed over it. Mrs. Buckland would set to work at it, and build it up to its original form with marvellous neatness. No mosaic work that I ever saw could have more time, trouble, or ingenuity than she has bestowed upon some of the specimens now at Oxford and elsewhere.

Among the Dean's papers I found the following characteristic letter from Dr. Wollaston, enclosing a receipt for this white cement for large and ponderous specimens. It runs as follows :—

"My dear Professor,—I send your great gun ship (along with your Rhinoceros tooth) sundry specimens cemented in my fashion. Try them, break them, do what you will with them, relying for full explanation on,

"Yours ever truly, H. Wollaston."

The cement mentioned above is thus made :—

1 part bees' wax,

4 parts resin,

5 parts powdered plaster of Paris.

Warm the edges of the specimens, and use the cement warm.

I have never used the cement myself, so that I cannot *guarantee* its efficacy. On the same paper are two more receipts for cement in my father's hand-writing. They may be useful, so I give them :—

CEMENT FOR MENDING SHELLS, USED AT PARIS.

Gum Arabic, two-thirds,

Sugar-candy, one-third,

White-lead.

The other is :—

A CHINA-MENDER'S RECEIPT FOR MENDING BROKEN CHINA.

Powder of Suffolk cheese dried, powdered, and sifted.

Unslacked lime in powder.

Equal parts. Mix with hard water, cold, on a slab till it is stringy. It is then fit for use. Put on each side of the broken edges, and hold before the fire till it sets, and in five hours it is done. N.B.—The more cheese the stronger it is. Boil the cheese to get the fat out.

Bones found in gravel pits, &c. are often in a very fragile state. In order to harden them, they should be washed frequently with a mixture of common glue and whitening; a little experience will indicate the proportions in which these materials should be used, there being always more glue than whitening.

I must not, however, fill up my book with receipts, many of which, better perhaps than the above, may be found in "The Encyclopædia of Practical Receipts," "Enquire Within upon Every Thing," and numerous other publications.

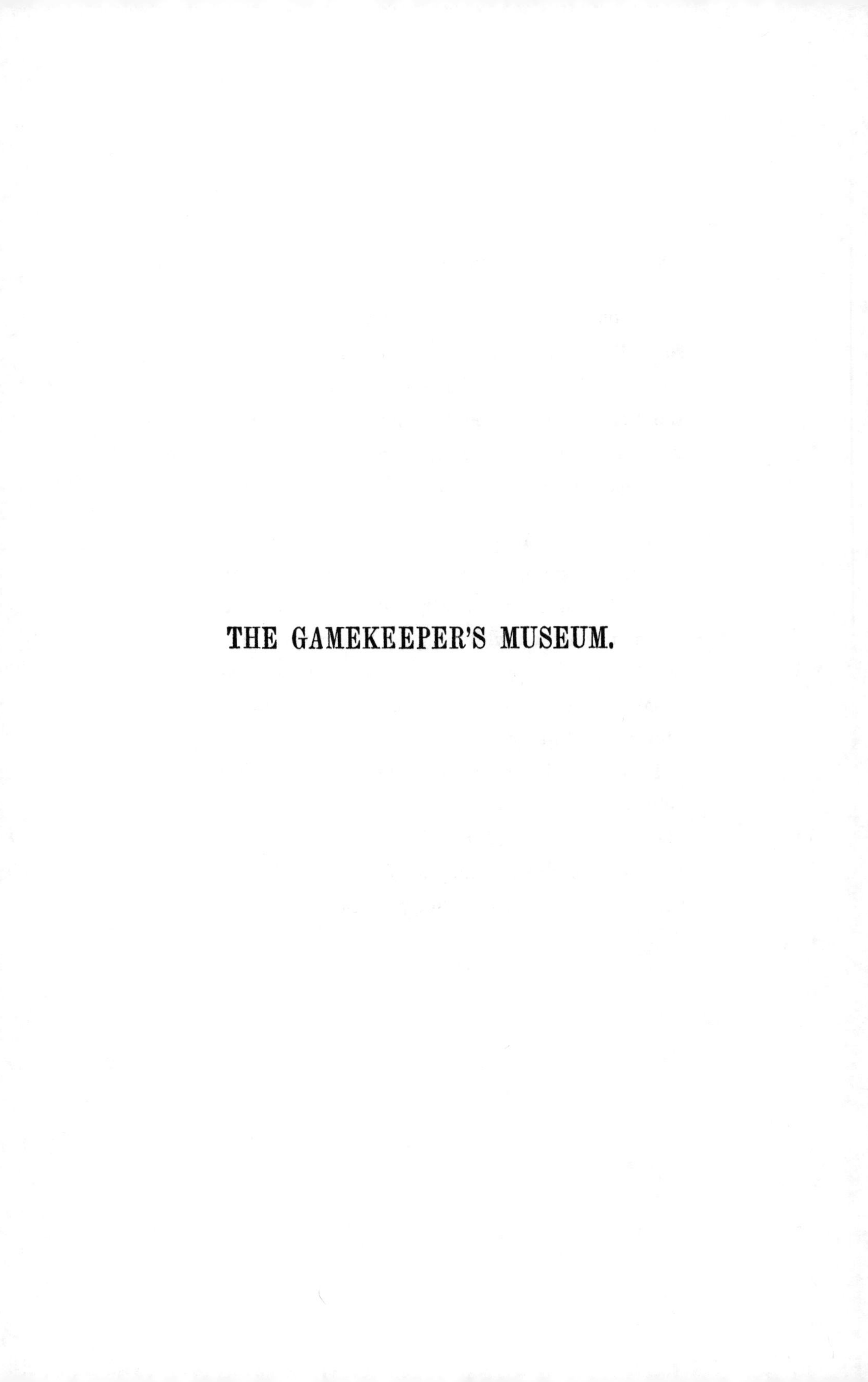

THE GAMEKEEPER'S MUSEUM.

THE GAMEKEEPER'S MUSEUM.

Much has been lately said and written upon educational subjects, and particularly upon the formation of local Museums.

There exist, however, unnoticed and despised, in most counties, local Museums, which will, if carefully examined, yield the most instructive information to those who take interest in the history of the animals now indigenous to England; animals whose representatives were co-existent with the ancient Britons; nay, many of them, as testified by the bones found in the caves of Kirkdale in Yorkshire, "flourished" (as the school-boy books have it) even in the days when Elephants, Bears,* Wolves,† Tigers, Hyænas and

* In the Field Newspaper I stated that a Gordon, in the year 1057, killed a Bear, and received permission of the king of Scotland of that day to quarter three Bears in his coat of arms. I have since received a kind letter from the

the great Rhinoceros roamed our broad acres, the undisputed inhabitants of this island.

As the cave bones are to the Geologist and the Palæontologist evidences of existence of the animals of what is generally called the antediluvian

Earl of Granard, who says:—"Your statement is incorrect; the shield, or rather the quartering of the Gordons, is three *Boar's*—not three Bear's-Heads; the three *Bears* on the shield or banner are the achievement of my family, one of whom killed a ferocious Bear in the Highlands, and was thus rewarded by Malcolm Canmore, or some sovereign about the same date."

† The Wolf was common in England at comparatively a modern date. King Edgar, A. D. 970, was a great Wolf-slayer, and attempted to get rid of them by setting criminals to work out their punishments at Wolf-hunting (*vice* oakum-picking of later times,) and petty offences were paid for in Wolves' tongues. "In the reign of Athelstan, Wolves abounded so much in Yorhshire, that a place of retreat was built at Flixton, near Scarborough, for the protection of passengers against their attacks. I have often longed to have a hunt about "Wolvesey" Palace, close to the College at Winchester. The ancient tribute of Wolves' heads was deposited at the gates of the palace, and they were probably buried close by, if one only knew whereabouts to look for them. Wolves began to multiply in the reign of Edward I. A. D., 1273, and one Peter Corbet was ordered to take measures for their extirpation in Gloucestershire. After this date there were no more Wolves seen in England and Wales, but they were not exterminated from North Britain till the seventeenth century. There are two claimants to the death of the last Wolf in Scotland, viz. Sir C. Cameron, 1680, and MacQueen, who died 1797. (See interesting paper, "Notes and Queries," November 12, 1859.) There is, in the Dublin Archæological Museum, a medallion on which is an engraving commemorative of the last Wolf killed in Ireland. I know not the date.

period, so does the "Gamekeeper's Museum" inform the naturalist that there are yet represen tatives of the ancient wild animals in existence in our woods and fields; and that although "man's hand has been against them" for many centuries, yet he has not utterly exterminated them from the face of the earth.

There is hardly an estate in England where the gamekeeper has not formed such a museum, or "larder." He selects generally the wall of a barn, or dog-kennel, or a tree near the house, as its site. He is the sole collector, preparer, and conservator of it. His business is to kill and destroy all the enemies of his special charge, the game. He does so, and nails or hangs up in his collection the heads of cats, hawks, owls, hedgehogs, stoats, weasels, and other so-called vermin, thereby demonstrating his power of field-craft, and his diligence in preserving his master's property.

After an absence of some months, perhaps during the whole of the nesting time and breeding season of the various kinds of game preserved in the coverts, it must indeed be satisfactory to the master to have ocular proofs of his gamekeeper's prowess, and such guarantees of future sport, offered by the increased number of specimens in the keeper's museum ; presuming always that the master is well assured of his keeper's honesty and good faith ; for whereas adulterations and frauds are unfortunately too common in sugar, coffee, pickles, &c., even so do we find that the stocking of the gamekeeper's museum is open to rascality, and is often replenished with victims not caught on the premises.

We well recollect the disgust depicted on the face of a well-known game preserver when he related how that his good-for-nothing gamekeeper had bought up all the cats in the neighboring town, cut off their heads, and nailed them up, as trophies of veritable captures in the neighboring woods ; and how that the same man had given a large order in an adjacent county for jays and magpies in order to swell the ranks of feathered culprits, and to palm off his purchased specimens as a *bona fide* collection shot and trapped by himself.

We remember the moral that the master drew and propounded to us, that it were better not to see the vermin at all, alive or dead; that is, they ought not to be alive about the coverts, nor yet be hung up when dead, as there is opening for rascality.

It requires a sportman of no moderate order to be able to go the rounds of his woods and game preserves, and from his personal knowledge of the art of woodcraft, to pronounce an opinion as to whether or not the game-keeper has done his duty in his absence, and whether the vermin that ought to be nailed up for the benefit of his game, as well as science, are still ranging at will or have been destroyed.

But lately I came across one of these rural museums, belonging to a friend of mine, on an estate not far from Brighton, where I had gone on a professional visit, and had half a day to look about me after attending to my patient. Three years it had taken the honest game-keeper to form his collection ; and not a little proud of it was he. The vermin, as he called them, "had lately got scarce," and all the heads and bodies hung up were perfectly hard and dry ; nor was there the slightest effluvium—that had passed off long ago—and the specimens were much in the same condition as the mummies we see in the British Museum, hard, dry, and grim. It is a curious thing that there seems to be an insect especially created to eat up dead animal matter.—Whether at home or abroad, I never found a dry body of an animal that had not, in and about it, specimens of a creature called "Dermestes lardarius ;" they are the "hoppers" of the ham and bacon merchants, and specially delight in eating dry animal flesh ; hence they are such plagues to the ham merchants. They are capital skeleton makers, and if the skins of the creatures in the gamekeeper's museum be removed, the skeletons will be found underneath in a most perfect state of preservation, and quite fit, after a little washing, for the cabinet. Hoppers also eat dry leather; and the late Mr. Baker, of Bridgewater, took advantage of their powers by setting them to work to make skeletons of delicate things, such as small birds, fish, frogs, lizards, &c.: neat workmen are these little hoppers, touching nothing but the flesh—and they clean much better than ants. The animal to be made into a skeleton should be soaked in water to get all the blood out, then dried and placed with the hoppers in a covered box; a few birds' feathers should be put over them, as they will work only in the dark. There were hoppers about even in the time of the Egyptians, and they used to eat the mummied bodies, for I have found them hard and dry inside the skull of a mummy in the Ashmŏlean Museum at Oxford.

The love of collecting trophies of the chase seems natural to man. In some instances they are held sacred, and the following is a good case in point. In his "Exploring Voyage up the River Kwora or Niger, (west coast of Africa,) in the Pleiad," Dr. W. Baikie writes: "In the centre of the village, (a long way up the river,) I found a pile of skulls and heads of hippopotami, buffaloes, deer, leopards and crocodiles; this part being considered sacred, and dedicated to the god of hunting. I wished to purchase some specimens, and after much debating, was allowed to offer terms. At Zhiru, where a similar heap was seen, the people obstinately refused to sell any of their spoil."

The victims in our gamekeeper's museum had not been nailed up by chance in the first vacant place, but arranged with a certain degree of taste, a row being apportioned to each species of animal.

The keeper's greatest enemies of course occupied the most prominent position; and in the top row no less than fifty-three cats' heads stared hideously down upon the visitor. There was a story attached to nearly each head: this cat was killed in such a wood, this in such a hedge-row, some in traps, some shot, some knocked on the head with a stick; but what was most remarkable was, the different expression of countenance observable in each individual head.

This one had died fighting bravely to the last; inch by inch it had yielded up its nine lives. Caught possibly in a trap in the early part of the evening by one of its legs, it had lingered the night through in agony, the pain of its entrapped limb causing it to make furious efforts to escape, and those very efforts adding additional torments to the wound. In the morning the keeper had come with his gun and his dogs; putting his foot on the spring of the trap, he had let out the wounded and exhausted animal to the mercy of his terriers; what little life was left in it the dogs worried out. It had died a martyr to its natural instinct.

Do you doubt this? Look at the head, now dried by the heat of two summers: the wrinkled forehead, the expanded eyelids, the glaring eye-balls, the whiskers extended their full stretch, the spiteful lips exposing the double row of tiger-like teeth envenomed by agony, teaches us all this. The hand of death has not been powerful enough to relax the muscles racked for so many hours with terror and pain.

Let us examine another head; what a difference in expression do we see in this cat at the end of the row; she has never been worried or tormented; stealthily creeping on the tips of her beautifully padded feet along some hedge-row, she has come within the range of the gun of the concealed keeper, and in an instant been *shot dead;* yes, shot dead; her calm look, her ears cocked well forward, the sagacious set of the muscles of her face remain to this moment,—so sudden was her death that other feelings had not time to work upon her expression and physiognomy. Her mummied head tells us the story of an unexpected and instantaneous death.

The cat has numerous muscles about her face, and she is capable of assuming nnmerous expressions. Let the reader nurse a cat on his lap, tickle her nose, ears, eyes, whiskers, &c., he will see what I mean. Above all, she cannot bear her whiskers to be touched or pulled; at the end of each of these stiff hairs is a large bnlb of nervous substance, which converts them into the most delicate feelers. They are of the greatest use to her when hunting about in the dark: in the Lion these nerve-bulbs at the end of the whiskers are as large as small peas.

There is yet another head in the museum from which we can read another history; it is that of a poor little puss who had died before she had attained the age of cathood. Her young life had probably been knocked out of her tender body with a stick; for her head still retains the playful look of the kitten; and there is a sort of a "what-have-I-done-look" about it, as though she had died with submission, and in ignorance of the keeper's anathema against her species.

I would remark that personally I have no antipathy to cats. I rather like them. I am now writing of them only in their character of vermin.

To the admirers of cats as pets, I can recommend a little book written about cats, and nothing but cats, viz: "The Cat, its History and Diseases," by the Honorable Lady Cust. 1856. Groombridge and Sons, Paternoster Row. Price 1s. Many persons dislike cats and other destructive animals because they are so blood-thirsty towards their fellow-brutes; but that one kind of beast should eat another, is a bountiful law of nature which I cannot now go into. The following verses have been kindly given to me; they were composed by Canning, after he had heard a discussion on the subject of animals killing and devouring one another.

Tell me, tell me, gentle Robin,
What is it sets thy breast a throbbing?
Is it that Grimalkin fell
Hath killed thy father or thy mother,
Thy sister or thy brother,
Or any other?
Tell me but that,
And I'll kill the cat.
But stay, little Robin, did *you* ever spare
A grub on the ground, or a fly in the air?
No, that you never did; I'll swear.
So I won't kill the cat,
That's flat!

But why have all these cats been killed? It is unfortunately impossible to convert woodland coverts into caged happy families, where cats will live in harmony with birds, and their co-existence in amity is incompatible with the law of nature. The master of our covert is jealous of the cat's destructive powers, and naturally applies the law of force to maintain his right to kill and destroy; hence this feline Golgotha. Let us hear what our friend the keeper has to say on the subject; he will tell us that the cat is the worst vermin in existence, for although not hungry, she will kill for sport, and if an old she-cat should lay up her young in the woods, it is incredible the amount of game and rabbits she will destroy.

"Prove the keeper's sweeping accusation against the feline race," says Pussy's friend. "Do they not prowl by night? How does the keeper know they do such mischief? Alas for the counsel for Pussy's defence! Let him go the rounds with the keeper in the morning, and under the warm shelter of a wall or bank, and even occasionally in the very middle of the rides and paths, shall he find the skins of fresh-killed rabbits turned completely inside out! a sure sign that the diner-out was of the feline genus. "And why might not the rabbit have been slain by a fox, weasel or other animal?" says Mr. Counsellor for the feline defendant. "Because," answers the keeper, "every animal has his own way of killing and eating his prey. The cat always turns the skin *inside out*, leaving the same reversed like a glove. The weasel and stoat will eat the brain and nibble about the head, and suck the blood. The fox will always leave the legs and hinder parts of a hare or rabbit; the dog tears his prey to pieces and eats it 'anyhow—all over the place;' the crows and magpies always peck at the eyes before they touch any part of the body."

Again, let the believer in the innocence of Mrs. Puss listen to the crow of the startled pheasant; he will hear him "tree," as the keeper calls it, and from his safe perch up in a branch again crow as if to summon his protector to his aid. No second summons does the keeper want; he at once runs to the spot, and there, stealing with erect ears, glaring eyes, and limbs collected together and at a high state of tension, ready for the fatal spring, he sees—What? the cat, of course, caught in the very attitude of premeditated poaching.

Again, let him listen to the tale of destruction, and learn how such and such a "nide" of young pheasants fell victims to his neighbor Turniptop's "Black Tom" in broad daylight, the "Black Tom" aforesaid, never, of course, having been known to poach before. Let him pay due deference to the keeper's opinion that to the cat alone he owes the loss of many a pheasant's nest. He is certain of this fact because the pheasant's eggs are all out of the nest, and scattered along in a line directly opposite to where the cat made her advance, the sitting bird having in her hurry and flutterings to escape swept her eggs out of the nest, for a foot or two along the track she had taken; just as if the reader should strike a pack of cards placed flat upon the table with his hand, driving them forwards in a line with his blow.

At Castle Forbes, in Ireland, last spring, the gamekeeper was in the habit of seeing every morning nearly a dozen young rabbits sunning themselves close under his window. Suddenly they disappeared. He went and looked round, and found the remains of one rabbit half pulled into a hole at the foot of an old tree; he instantly put a trap down, and caught successively an old she-cat and four young ones. The old she-cat he recognised as the cat belonging to the kitchen at the castle, which had been missing for a week or two previously.

It is unfortunate for cats in general, that if one of their race once takes to poaching, their nature prompts them to continue these evil ways. They find out that game is better eating than rats and mice; they leave the homestead and take to the woods; thenceforward becoming perfectly useless in the domestic economy of the farmer or cottager. So strong is this passion for hunting when once acquired, that it is impossible to break them of it. We once knew of a cat against which sentence of death had been recorded, but the owner begged its life on condition that it should be shut up every night and well fed. The very first night of its incarceration, it escaped up the chimney, and the following morning astonished the eyes of the gamekeeper as the soot-begrimed occupier of one of his traps. Feed the cat as much as you please, make a pet of it, &c., you will never break it from night hunting, though you may succeed in stopping its *diurnal* rambles.

It is quite wonderful to see a cat jump down heights. She never seems to hurt herself, or to feel giddy with the fall; she always falls on her feet, and these are so beautifully padded that they seldom or never get brokeu. I never knew of a cat breaking its leg from an accident, but in one instance, and that was a *French* cat, which fell down stairs in the most stupid manner. Why does not the cat get a headache after her deep jumps?—why does not she get concussion of the brain, as a man or a dog would, if he performed a similar acrobatic feat? If we take down one of our dry cat's heads off the keeper's museum wall, and break it up, we shall see that it has a regular partition wall projecting from its sides, a good way inwards, towards the centre, so as to prevent the brain from suffering concussion. This is, indeed, a beautiful contrivance, and shows an admirable internal structure, made in wonderful conformity with external form and nocturnal habits. Apropos to feline skulls, I may here give the lesson I learnt from Professor Quekett, of the Royal College of Surgeons, as to how to tell a Lion's skull from a Tiger's skull; for, when placed side by side, there is not much perceptible difference between them. All that is required is, to remember that the word "Level," as well as the word Lion, begins with the letter L. In the lion's skull the points of the four bones, which form the nose part of the face, are *on a level* with each other. In the tiger's skull, on the contrary, the points of the two central bones run higher up the skull than the two outside ones.

It has been said, with partial truth, that cutting off the ears of cats to the level of their heads, and at the same time removing with scissors the hair all round the exposed aperture of the ear, will keep cats out of the woods: for the simple reason that, being earless, when they go out hunting among the wet bushes and grass, drops of water will get into the internal cavity of the ear, the effect of which is, as we ourselves know, to cause a disagreeable sensation almost amounting to pain. Let the reader examine the cat's ear as she is purring before the fire; he will find that the opening to it is amazingly large, though partially concealed by hairs which grow up from its internal surface. The cat, from experience, understands the doctrine of cause and effect, and stays at home when her ears are cropped. But does she always stay at home? No; the cunning creature waits for a nice fine day, when the grass and hedges are quite dry, and off she goes at her old poaching tricks. Put the intelligent sportsman in the witness-box; he informs us that he has often shot or trapped "doctored" cats in the heart of the covert in fine dry weather, but in wet weather, always those who have never been submitted to this mutilation. A cat hates nothing so much as getting wet, and hence we learn from "High Elms," a good plan to catch a poaching cat. He says, "In damp weather, or when the dew is rising, a cat almost invariably walks along a wall to get to her hunting grounds. I have caught many a one by this means; I need hardly tell you no bait is required." Again, "If there should happen to be a watercourse running through the plantation, a trap set in a coping, where the stream goes under the wall, will be very effective as a cat will be at the trouble of climbing on to the wall, crossing over the stream, and jumping down again, sooner than leap over the stream itself." No vermin is more easily trapped than the cat; in summer, when rabbit-paunches will not last, but get full of fly-blows, a little valerian root will serve equally well; they smell it afar off, and eagerly come near to rub themselves on it, though the pleasure they can derive from it is to us bipeds unintelligible.

A facetious young urchin, at home for the Christmas holidays, knowing well the love cats had for valerian, once played an old lady a pretty trick. He put some of this plant under the hearth-rug one evening; puss soon found it out, and began scratching and rubbing her back upon it, and then getting up and dancing about, till the poor old lady got frightened, thinking the cat was suddenly possessed. The valerian was quietly taken away, and puss recovered her self-possession, which confirmed the old lady in her original opinion.

Where cats are numerous, a hutch-trap, such as are used in warrens, will be found very killing; it must be baited with fish of some kind, but a red-herring is most captivating. If another red-herring be dragged about the covert, and then up to the trap, puss will run up the drag and of course be trapped, provided always she is hungry and inclined for a fish dinner. A caged cat is an awkward customer to handle; so that unless the cat-catcher knows how to kill his victim, he had better not have set the trap at all. Let him again beware, when he comes to look at his trap, not to be too eager or curious to open the door to see what he has caught; for the cat has an unpleasant habit of striking at the human face divine the moment she sees light, and her sharp claws and pointed teeth can make severe wounds, difficult to heal. There is only one safe way of getting the cat out of the trap, and that is to place a sack over the door-end of the trap, and then rattle the other end with a stick; the foolish creature goes into the sack immediately, and you can then let the "cat out of the bag" at the proper time and place.

Should the hutch-traps be made with lattice-work or grating, a very simple method of killing the "betrayed one," is to insert a wire noose between the bars, over the victim's head, and so, by strangulation, put a speedy end to its misery.

Not a single head of a genuine wild cat did we see in the museum—they are now nearly extinct; all the heads were those of house cats. "The wild-cat is the only species of the family which is indigenous to the British Isles. It is now almost entirely restricted to Scotland, some of the woods north of England, the woody mountains of Wales, and some parts of Ireland. It is necessary to guard against confounding the wild cat with numerous instances of escaped domestic cats, returning to a state of almost absolute wildness, breeding in the woods, and feeding on birds and quadrupeds. The assertion that the wild and domestic cat will breed together, I believe absolutely to be without foundation. The head of the wild cat is triangular, strongly marked; the ears rather large, long pointed and triangular; the body strong and rather more robust than that of the domestic cat. The tail of equal size throughout its length, or rather larger towards its extremity."—*Extract from "Bell's Quadrupeds.*" I learnt, when a little boy, from my father, that the common cat was brought over here from Italy by the Romans; and that a wild cat's tail ends in a tuft, as if it had been chopped off with a hatchet on a block; whereas the common cat's tail ends in a point, and tapers to a point as if it had been gradually drawn out of the body.

"At the village of Barnborough, in Yorkshire, there is a tradition extant, of a serious conflict that once took place between a man and a wild cat. The inhabitants assert that the fight commenced in an adjacent wood, and that it was continued from thence to the porch of the church, where each died of the wounds received. A rude painting in the church commemorates the event; and the red tinge of some of the stones (though probably natural) has been construed into bloody stains, which all the soap and water hitherto used, have been unable to efface.

In the time of Richard II., 1377, wild cats were reckoned among the beasts of the chase, and there was an order that no abbess or nun should use more costly apparel than that made of lambs' or cats' skins. What would the Abbesses have said to the beautiful North American Sea-otter's skin, now worn by English ladies?

But a few minutes after my inspection of the museum, I saw a house cat out in the open fields, creeping, as only a cat *can* creep, up a hedgerow ; evidently she had not seen the Golgotha of her species, or she would have remained home, warming herself by the fire. Many a poor old woman has wondered, as only an old woman *can* wonder what has become of "our Puss?" The keeper could, if he liked, answer her question, pointing triumphantly to his museum : "There she is on my barn-door, executed for a poacher." But in all probability the old lady never once divined that the reason why her poor puss slept so soundly all day, and dozed so lazily in the window, was because she had passed the previous night in the game preserves.

The heads of dogs are seldom seen in the keeper's museum. He generally buries them. I have heard a theory that the reason why the gamekeeper generally can produce finer gooseberries, cabbages, &c. than his neighbors, is that his garden is well manured with defunct dogs buried all about it. If an Englishman is persecuted and followed by a yelping cur, he can generally manage to get rid of him by stooping down and pretending to pick up a stone, for all curs have a mortal dread of a thrown stone; but on the *bogs* of Ireland the dogs don't care a bit if the person they are barking at pretends to pick up a stone; they know, cunning brutes, there are no stones on the bogs to be picked up and thrown at them, but they act very differently if there happens to be a heap of stones anywhere handy. It is an unpleasant situation to be attacked by a dog; if you are so circumstanced, never attempt to run, try throwing a stone at him, or present your hat in your hand, and when he has seized it, hit him with a stick across the *nose* or *fore leg*. These are the most vulnerable points in a dog; a blow on any other part of the head but the nose won't hurt him a bit.

If a dog comes up to you and growls, and won't be friendly, don't withdraw from him; but put on a bold face, and stretch your hand towards him, keeping it quite still (if you withdraw it after stretching it out he will bite you); the dog will come up and smell the hand, and, having once done this, will be your friend for life. A chimney-sweep once made a match to fight a bull-dog single handed, armed only with his brush. He entered the arena with his brush in one hand and a foot of bramble bush covered with thorns in the other. The dog sprung at him; he presented the bramble bush to the animal, who seized it in his mouth, and so got hooked by the thorns on it; the chimney-sweep belabored him over the head and nose with the back of the brush, and won the match. We may learn from this, that if a man is attacked by a bull-dog, he should hold out a stick between his hands, and present it to the dog who will seize it, and give the man time for further measures. A rat-catcher lately told me that he had a monkey that would be "a match for any dog in any pit." The monkey was given a short stout stick; he watched his opportunity, sprang on the dog's back, it was impossible for the dog to throw him, and the monkey beat him about the head at his will.

Dogs, like cats, sometimes run wild, and nothing is more difficult to kill than a wild Irish dog, hunting by himself, in the bogs or on the open ground. He will always keep well out of shot, and really there is only one way to get near him; and that, it must be said, is a gross imposition on canine confidence. When the dog is seen hunting about, take no notice of him, but pretend to hunt about also yourself; beat the bushes, and cheer lustily: the unsuspicious dog, prompted by his instinct to be of use to man, comes to afford assistance; he is put off his guard, approaches within shot, and is carried off defunct by the keeper to be converted into gooseberries.

Shooting poaching-dogs often gets gamekeepers and others into hot water with the owners of the dogs. I lately heard of a gamekeeper who shot a farmer's dog which was always in and out of the coverts, doing harm to the game. At last, one day the keeper met the dog wandering along, off his guard, and shot him; he immediately dug a hole and buried him. While thus occupied, he happened to look up, and there he saw the dog's master watching him through the hedge some distance off. The farmer did not come nearer, but walked away quickly in the direction of the mansion-house of the keeper's master. He was ushered in, and made his complaint that he had seen the keeper shoot his dog. The master sent his steward off to the gamekeeper's house, with orders to inquire about it, and to dismiss the keeper if the charge were true.

But the proof was wanting, and the farmer proposed going to dig up the dog then and there, so off they all started, and dug away the freshly moved earth till they exposed the dog's tail: when they dragged the body out, what was the farmer's bewilderment to find it was not *his* dog, but a dog he had not seen before. Now the keeper had really shot the farmer's dog and buried him in this place; and having plenty of ready wit, reasoned that the farmer, having seen the deed, had gone off to make a complaint against him. He therefore ran off home as quickly as he could, shot a cur he happened to have, that luckily resembled the farmer's dog, put him in his great hare-pocket, and buried him in the hole he had just dug, taking out, of course, the farmer's dog he had shot. The poor farmer was not up to this quick manœuvre, rubbed his eyes, saying it was very odd, he could not make it out, he was certain he saw *his* dog buried not half-an-hour ago in that very hole, but it was not there now. The matter ended by the farmer accompanying his apologies to the keeper with a fee; and to this day it is a mystery in the parish as to what became of the dog.

In former times the rangers of the New Forest were very particular about poaching-dogs. I lately spent a couple of hours at Lyndhurst; the most considerable village within the precincts of the Forest, of which it is in reality the capital town. In the centre of the village or town stands the King's House. "The ancient hall of the hunting-seat has obtained the name of Rufus's Hall. Here the Forest Courts are held. The seats at the upper end of the 'verderers' and other officers of the Courts are all of sturdy oak and of great age, as are also some capacious tables. The walls are hung with antlers of stags, and with the skins of two eagles, which were shot many years ago by some of the keepers of the Forest." In Rufus's Hall I was shown what in modern parlance might be called a "dog guage," to measure dogs. Over the fireplace was hanging a curiously-shaped bit of iron, which is reported to have been the stirrup ot William Rufus. There are traces of gold upon it. Its extreme width at the bottom is ten inches and a half, its depth seven inches and a half, measured all round, it compasses no less than two feet seven inches. It was formerly applied, so says the legend about it, as a test for ascertaining what dogs kept within the Forest should "suffer expeditation." If a dog could not be drawn through the stirrup, he was to suffer this operation to disqualify him from the pursuit of deer. The above process is thus described in Manwoods on Forest Laws, p. 35:—"The way of expeditating mastiffs is done after this manner, viz. three claws of the fore-foot shall be cut off by the setting one of his fore feet upon a piece of wood eight inches thick and a foot square; with a mallet, setting a chisel of two inches broad upon the three claws of his fore feet and at one blow cutting them clean off."*

* I am much indebted to Mrs. Fenwick of Lyndhurst for a transcript of these particulars.

When my uncle, T. Morland, Esq. was master of the old Berkshire hounds at Sheepstead, near Abingdon, Berks, I used to help the huntsman clip the hounds' ears into a rounded shape, just in the manner above described. We found it always much more difficult to do the *second* ear than the first, the hound being naturally unwilling to come up to the block the second time; it also required some management to make the two ears both of a size. Nobody who has not tried it can tell how difficult it is to cut the ears of common terriers so as to make them cock up nicely. There are men in London who make a business of ear-clipping.

Underneath the row of cats' heads in the gamekeeper's museum, were deposited dry bodies of the birds, the feathered enemies to the game. there were magpies, jays, owls, crows and hawks —of the latter, I counted thirteen bodies, principally sparrow-hawks—and kestrels. The keeper called these latter "Fanner Hawks," and a very good name this is, for they swim, so to say, about in the air, surveying with telescopic eye* the ground for their food; having discovered it, they remain hovering over it, suspending themselves in the air by means of a rapid movement of their wings, much like a bee when inspecting the inside of a flower. In other parts of England they call these birds "wind hovers;" an equally good name, inasmuch as it indicates the habits of the animal as it hovers in the wind, previously to making its arrow-like descent upon the unconscious mouse below.

A great observer of nature, and a clever sportsman told me that hawks have their regular beat, and frequent daily the same line of country, soaring along for miles and miles in quest of prey. So strongly impressed was he with this idea, that he always marked the time and place

* In the eyes of these dried hawks the peculiar construction of the hard coat of the eye, the sclerotic, can easily be seen; it is formed of numerous plates beautifully fitted together, in a manner which a human workman could hardly equal, and certainly could not excel. It has its special uses, which I cannot here enter into. Many of the ancient Saurians of former days had the sclerotic coats of their eyes formed after this plan See Dr. Buckland's "Bridgewater Treatise," Description and Plates.

when he saw a hawk on the hunt, and sure enough the next day would find my friend at the spot, waiting in ambush gun in hand, and consulting his watch, as confidently as if he were expecting a friend by the most punctual of railways. He assured me that he always found the hawk true to his time by half an hour or so, and seldom vary his line of flight by more than a hundred yards.

I saw no heads, or remains of the fork-tailed kite in the gamekeeper's museum; it has become nearly extinct in the southern parts of England. When my father was a young man at Oxford, about 1803, these birds were numerous in Bagley Wood, and he frequently saw them sailing about over the Thames, when walking round Christ Church meadow.

I think there were one or two raven's heads nailed up; these birds are also getting very scarce. I see young ones in Leadenhall Market for sale every year. Last year I bought two; one I sent to a friend, a Prefect at Winchester

School, the other is become the "Regimental Raven," and is now hopping about the barrack yard; he is a capital rat and mouse catcher. I gave ten shillings each for them.

The hawks had been killed by the keeper, because they destroy the young pheasants and partridges; the crows, jays, and magpies, because they destroy the eggs of the game. If we examine their hard, sharp-pointed, conical bills, we shall see at once what a capital instrument Nature has given these birds, to enable them to fill the position allotted to them in the chain of animal economy. It is a compound of a dagger, a bayonet, and a club; there is no egg hard enough to withstand one peck from its point; there is no rabbit's hide that is proof against its bayonet thrust; nor is any mouse alive after one or two sharp blows, when the bird uses it as a club. I lately turned out a mouse to a tame magpie—the poor little thing ran for his life towards the nearest cover; but mag's sharp eye had seen it, and a few hops soon brought her up to the mouse, just as he was getting under a board. She caught him in her bill, and threw him back again into the open; she then gave him a stab, which made a cripple of him, and knocked out of his tiny panting body with a few blows of her dagger-like beak the little life that remained.

Magpies, if properly trained, can be taught to do the work of retriever dogs, on a small scale. Mr. Blick, of Islip, is in the habit of shooting sparrows in his garden: on these expeditions he is always accompanied by his dog and his cat, who run round him in great delight whenever they see the gun taken down. At the same time, out hops from under the bushes where he has been hunting for worms and slugs, a pert impudent-looking magpie, jerking his long tail, and croaking "mag, mag," out with ten magpie power. A shot is fired at the unsuspecting sparrows, who are filling their little crops with the corn spread out to delude them into the idea that they are welcome visitors to the yard. A shot is fired, a victim falls lifeless to the ground; up rush dog cat, and magpie, each anxious first to secure for themselves a dainty morsel. It is a good race, but the magpie generally gets in first, and seizing the panting bird, hops off with it underneath the dense shrubs, closely pursued by the dog and the cat, who are obliged to look on patiently at their more successful competitor the magpie, who is now picking off the sparrow's feathers and throwing them down, as if in mockery, on the heads of his rivals, the said rivals being unable to reach him, his natural sagacity having suggested to him the propriety of taking up his position on a twig, just too high for the dog to reach by jumping, and too slender to bear the weight of poor disappointed puss.

Magpies have always been connected with some superstitious stories; even the peasants of Norway say that they have to do with witches: and who does not know the old rhyme about magpies so often quoted by good folks setting out on pic-nic expeditions? A magpie appears alone, or in company, and immediately somebody says :—

> One the sign of sorrow,
> Two the sign of mirth,
> Three the sign of a wedding,
> Four the sign of a birth.

I do not know what the French peasants think of magpies, for all along the railway from Boulogne very nearly into Paris, I saw a magpie's nest in almost every tree—their numbers in this case certainly proved prophetic, for I was on the road to a wedding at Paris. It is often said, that it is unlucky to rob a magpie's nest; it is a fact, that when this has been done, the parent-birds become more destructive to the hen-wife's poultry than they were before. Magpies are often taught to talk, and it is very curious that they pick up the accent of their teacher. Outside the cottage of a Berkshire villager, I espied a fine magpie in a cage, and he began talking away in as broad Berkshire as ever I heard. I also recollect a German student, who said he could talk English; he certainly could say a few words, but he spoke with a broad Yorkshire accent; he had picked up the accent from a Yorkshireman, his fellow-student. I may here say, that it is no use whatever, as is often supposed, to split the tongues of these birds in order to give them facility of speech. It is cruel work, and does no good. Magpies can talk sometimes even better than men. I was told of a conceited young gentleman who naturally stammered, coming up to the owner of a magpie who was a working man, and after rattling the bars of the cage with his gold-headed cane, he said, "I say, my man, can y-o-u-r mag-mag-mag-pie t-t-t-talk?" "Yes," said the man, "a precious deal better than you can, or I would wring his neck on the spot."

As the museum was situated near the sea-coast, I was therefore not surprised to see in the collection a Royston, or hooded crow. This bird's proper home is the sea-shore, where his business is to follow the retiring tide, and eat what is left thereby. Nor does he object to small crabs, and those curious sea anemones which the good folks at Guernsey so aptly call "bloody fingers." Having capital wings, he often takes a look at the rocks, where the gulls and other sea-birds build their nests and place their eggs. When these fail him, he will take an inland journey, and very naturally mistakes a pheasant's egg for a gull's egg. The keeper, in his turn, very naturally seeing what he is after, mistakes him for a carrion crow, shoots and gibbets him—hence his appearance in the museum. The keeper calls him the saddle-back crow ; a good name again, for his head, tail, and wings are black, and the rest of his body of a fine ash-grey color, so that he looks very like a common crow with a saddle on his back. Our French neighbors too, whose shores he also visits, have evidently, with the same idea, christened him *Corneille mantelée*, or crow with a cloak on. These crows are very quick in finding out dead or wounded birds. A great sportsman tells me that he has often gone at daylight, to pick up wild fowl which he had shot the previous evening, and found that these saddle-back crows had anticipated him, and made a meal of his wild ducks and teal.

In Ireland these birds are called "Scalcrows," and they are very numerous, frequenting chiefly the vicinity of the larger lakes and rivers. Having never had an opportunity of comparing the Royston crow of England with the Scalcrow of Ireland, though it is most likely that they are the same species under a different name, I cannot pronounce an opinion why in the sister country they appear to be far more destructive and formidable enemies to game than they are in England. In Ireland a pair of scalcrows will hunt a moor or bog, and quarter the ground as regularly as a brace of well-broken pointers—and woe to any brood of grouse or young leveret they may find. Nothing comes amiss to them. The old hen-grouse in vain sits close on her nest. The scalcrows persecute her, and hover round and round about her, alight, and fairly beat her off their luxurious banquet of egg omelet. On an Irish bog they are without exception the worst vermin the gamekeeper has tc contend with.

Even in England, game-eating and egg-destroying birds are very wary, and difficult to get near with a gun. The best way, therefore, of ridding the covert of them, is to poison a dead rabbit and place it in their way; they quickly find it out, peck at it, and suffer death in consequence. At Castle Forbes, the keeper has picked up as many as twenty-one magpies and crows to one rabbit at one time, and seven magpies and seven carrion crows at one another; but this did not last long; somehow or other the cunning birds found out that it was dangerous to peck at dead rabbits; in vain therefore were they laid down; the crows and magpies were for a season triumphant. But their enemy, man, was more cunning than they; he shot some wood-pigeons, poisoned them, down came the less cunning birds, and not suspecting treachery in a wood-pigeon, though they knew it was present in a rabbit, they pecked and died. Strychnine,* the most deadly poison made use of, is mostly sent to Australia, where it is used by the colonists for killing the dingoes, or native wild dogs, also the eagles, hawks, and vermin of that wonderful continent. The colonists have remarked the peculiar effects of this poison on the creatures who eat it. An eagle has been seen to peck at the poisoned bait and then soar away; in the midst of his flight the poison has taken effect, the bird falls swiftly to the ground, and, in a very short time, the body becomes perfectly rigid and set, preserving, when dead on the ground, the beautiful attitude of flight peculiar to this splendid bird.

I saw no trace of an Eagle in the Museum, but still these birds sometimes come very far south. When we were quartered at Windsor, in 1856,

* At the request of the editor of *The Field*, I have lately made a series of experiments relative to a dispute as regards the action of strychnine. I arrived at some interesting results, which will be found in that publication should the reader take interest in these matters.

an eagle was shot in Windsor Forest, and I then sent the following account to "Household Words."

"The Royal Forest of Windsor has lately been honored by a visit from a royal bird. The Eagle of the north visited the domains of the Queen of the south. The particulars are as follows:—

"On the afternoon of the 12th of December last, as one of the officers of the garrison of Windsor was riding in the great park, not far from the statue of King George III., at the end of the Long Walk, he was surprised to see a large bird on the ground, gorging himself with a rabbit. He advanced towards it, but the bird flew up into a tree. When on the tree it appeared to have a chain round its leg, but this was afterwards ascertained to be a portion of the rabbit it had just been eating. The pursuer having clearly made out that this large bird was an eagle, a most unusual visitor to the Royal Forest, rode off immediately to the keeper's lodge with the news. The keeper, while mounting his pony, stated that the bird had been seen about the forest for four or five days, but had always kept out of shot. When they both got back to the place where the bird was sitting, the keeper concealed himself and his gun, while the officer rode round the bird, endeavoring to drive him over the ambush. Off he went at last, but flew wide of the keeper. Then came the riding part of the business, partaking more of the character of a steeplechase than of hunting. By dint of hard and difficult galloping among rabbit holes, thick ferns, and open drains, the eagle was again marked down in a clump of trees. Then followed a little stalking. The keeper on his pony and his companion on his horse advanced carefully, but the cunning bird would not allow them to come near. The keeper then get off his pony, and walked alongside the horse, which was of a grey color, and seemed not to alarm the bird so much as the pony, which was of a dark color.

" After a few steps, the keeper suddenly and quietly glided behind a tree, and the grey horse and his rider advanced farther. To divert the attention of the suspicious bird, the latter wisely made as much noise as he could, tapping the saddle with his whip, riding among the thick ferns, and pretending all the careless unconcern he could assume. In the meantime the keeper got near, and fired both barrels. The bird flew away, but had been evidently hard hit, for his flight was labored and near the ground. He alighted at last upon the bough of a young tree, where his drooping wings and fainting form made him look more like an old coat hung up as a scare-crow than an eagle.

"Both the pursuers then rode up, and again, although wounded and bleeding, the courageous bird started off, but he could not go far: it was his last flight; for, in another minute, he dropped dead, shot through the right eye. The former shot had hit him in the body, but had in no way damaged his plumage. Shortly afterwards we inspected this noble bird, and found him to be a fine specimen of the white-tailed Sea-eagle. He measured with ontspread wings eight feet; the length of his body from his beak to his tail was three feet two inches; and he weighed twenty-two pounds. From his plumage, which was in excellent condition, it seemed probable that he was a wild bird; there being no marks either of cage or chain to indicate that he had ever been in captivity. His skin has been well preserved in a well-chosen attitude. Three or four years ago a Golden-eagle was shot in the forest, and presented by his Royal Highness the Prince Consort to Eton College."

When at Oxford, I had a tame Eagle, which I kept outside my rooms in "Fell's Buildings," at Christ Church; and my Eagle got me into much trouble, as it was an un-academical bird. I was obliged to banish him to the care of Mr. Osman, the bird stuffer, in St. Aldate's Street. When I left Oxford, I took the eagle with me, and, having received a University education, I suppose he thought he must do something in this great metropolis. So he took unto himself wings, and flew; and that upon a memorable day in the annals of Londoners. An account of his performances at the time was written by me, and published by a friend of mine;* it ran as follows:—

"The Chartists and special constables of Westminster, who were preparing on the 9th of April, 1848, for the grand 'demonstration' of the following day, beheld with varied feelings an omen which they interpreted according to their views. A magnificent eagle suddenly appeared sailing over the towers of Westminster Abbey; and, after performing numerous gyrations, was seen to perch upon the summit of one of the pinnacles. He formed a most striking object, and a crowd speedily collected to behold this unusual spectacle. After gazing about him for a time he rose, and began ascending, by successive circles, to an immense height; and then floated off to the north of London, occasionally giving a gentle flap with his wings, but otherwise appearing to sail away to the clouds, among which he was ultimately lost.

* See Zoological Notes and Anecdotes. Bentley, New Burlington Street, 1852.

"Whence came this royal bird, and whither did he wend his way?

"His history was as follows. Early in 1848 a white-tailed sea eagle was brought to London in a Scotch steamer, cooped up in a crib used for wine bottles, and presenting a most melancholy and forlorn appearance. A gentleman, seeing him in this woeful plight, took pity on him, purchased him, and took him to Oxford; he being duly labeled at the Great Western Station, 'Passenger's Luggage.' By the care of his new master, Mr. Francis Buckland, the bird soon regained his natural noble aspect; delighting especially to dip and wash in a pan of water, then sitting on his perch with his magnificent wings expanded to their full extent, basking in the sun, his head always turned towards that luminary, whose glare he did not mind. A few nights after his arrival at his new abode, the whole house was aroused by cries as of a child in mortal agony. The night was intensely dark, but at length the boldest of the family ventured out to see what was the matter. In the middle of the grass-plot was the eagle, who had evidently a victim over which he was cowering with outspread wings, croaking a hoarse defiance to the intruder upon his nocturnal banquet. On lights being brought, he hopped off, with his prey in one claw, to a dark corner; where he was left to enjoy it in peace, since it was evidently not, as at first feared, an infant rustic from the neighboring village. The mystery was not, however, cleared up for three days; when a large lump of hedgehog's bristles and bones, rejected by the bird, at once explained the nature of his meal. He had doubtless caught the unlucky hedge-pig (as it is called in Oxfordshire) when on his rounds in search of food; and, in spite of his formidable armor of bristles, had managed to uncoil him with his sharp bill, and to devour him. How the prickles found their way down his throat, is best known to himself: but it must have been rather a stimulating feast.

"This eagle was, with good reason, the terror of all the other pets of the house. On one occasion he pursued a little black and tan terrier, hopping with fearful jumps, assisted by his wings, which, happily for the affrighted dog, had been recently clipped. To this the little favorite owed his life, as he crept through a hedge which his assailant could not fly over; but it was a very near thing, for if the dog's tail had not been between his legs, it would certainly have been seized by the claw which was thrust after him just as he bolted through the briars. Less fortunate was a beautiful little kitten, the pet of the nursery,—a few tufts of fur alone marked the depository of her remains. Several Guinea pigs and sundry hungry cats too paid the debt of nature through his means; but a sad loss was that of a jackdaw of remarkable colloquial powers and unbounded assurance, who, rashly paying a visit of a friendly nature to the eagle, was instantly devoured. Master Jacko, the monkey, on one occasion only saved his dear life by swiftness of foot, getting on the branch of a tree just as the eagle came rushing to its foot with outspread wings and open beak. The legend is, that

Jacko became suddenly grey immediately after this; but the matter is open to doubt.

"One fine summer's morning the window of the breakfast-room was thrown open, previous to the appearance of the family. On the table was placed a ham of remarkable flavor and general popularity, fully meriting the high encomiums which had been passed upon it the previous day. The rustling of female garments was heard—the breakfast-room door opened, and—oh, gracious! what a sight! There was the eagle perched upon the ham, tearing away at it with unbounded appetite, his talons firmly fixed in the rich deep fat. Finding himself disturbed, he endeavored to fly off with the prize, and made a sad clatter with it among the cups and saucers; finding, however, that it was too heavy for him, he suddenly dropped it on the carpet, snatched up a cold partridge, and made a hasty exit through the window, well satisfied with his foraging expedition. The ham, however, was left in too deplorable a state to bear description. The eagle was afterwards taken to London and placed in a court-yard near Westminster Abbey, where he resided in solitary majesty. It was from thence he made his escape on the 9th of April. He first managed to flutter up to the top of the wall, thence he took flight unsteadily, and with difficulty, until he had cleared the houses, but as he ascended into mid-air his strength returned, and he soared majestically up as has been narrated. After his disappearance his worthy master said with a disconsolate air: 'Well, I've seen the last of my eagle!' but thinking that he might possibly find his way back to his old haunt, a chicken was tied to a stick in the court-yard, and, just before dark, master eagle came back, his huge wings rustling in the air: the chicken cowered down to the ground, but in vain—the eagle saw him, and pounced down in a moment in his old abode. Whilst he was busily engaged in devouring the chicken, a plaid was thrown over his head and he was easily secured. After this escapade he was sent to the Zoological Gardens, Regent's Park, where he may be recognised by having lost the outside claw of the left foot."

I often had fights with this bird, who was of rather a savage disposition. He once got hold of my leg, and his claws were obliged to be unfastened one by one. It is very easy to handle an eagle, hawk, owl, or other such bird, with sharp talons, if you only know how. The brute always strikes first with his claws, and then pecks with his bill. Remembering this, allow the eagle, hawk, or owl, to clutch something or other, say a broom-handle, or a walking-stick; then quickly throw a Scotch plaid, or a blanket, over his head, when you may release the stick, let him clutch a bit of the plaid, tie his legs, and he can be carried anywhere. In this manner I once carried my eagle on the top of an Atlas omnibus, from the Zoological Gardens, Regent's Park, down to Westminster Bridge. He got his head out and pecked me in Regent Street, but I soon secured him again. He was furious at being treated in this way; he ruffled up his feathers and looked spitefully at me when I untied his legs and let him loose again, after he had arrived at his destination; but I made friends with him by means of a good lump of beef-steak.

In the present stage of man's history, birds are not often consulted, under the impression that they have the power of foretelling events; but in the days of the Romans, the art of augury* was at its height; no Gamekeepers' Museums then existed, and instead of being persecuted, the wild birds were consulted and looked up to with the same sort of reverence as is the surgeon by the patient after he has examined his bad knee, and before he gives his opinion.

In the summer of 1844, I accompanied my father to Weymouth: my usual office on these occasions (as I have described in my memoir of him in the third edition of his "Bridgewater Treatise") was to carry his inseparable bag, which he used for collecting. When at Wey-

* "Augury, or divination, or soothsaying, by the flight or singing of birds. Qu. Avigerium vel Avigarrium."—Old Dictionary. "Da, pater, augurium; atque animis illabere nostris."—Æn. iii. 89.

mouth we visited the remains of a Roman temple, and a Roman villa, found on the top of Jordan Hill, close by, by Mr. Medhurst of Weymouth. In these Roman remains, among other things we found bones of animals and birds, that had been used for the purposes of sacrifice and augury; and among them those of birds which have since been deposed from their sacred offices, and classified by us moderns under the ignominious term of vermin. *Tempora mutantur*, and so much the worse for the birds.

My father afterwards read a paper on this subject, which I will not spoil by mutilation,* as it contains much that is very interesting; it runs as follows:—

"Dr. Buckland then proceeded to give a detailed account of the remains of many Roman buildings, discovered recently by Mr. Medhurst, near Weymouth. The neighborhood abounds with vestiges of Roman occupation. The large military station and Roman walls, Roman camp,

* See Proceedings of Ashmolean Society at Oxford, Vol. I. p. 55, Nov. 4, 1844.

and amphitheatre at Dorchester, contiguous to the gigantic British triple camp of Maiden Castle, are well known. The situation of Weymouth Bay and Weymouth Harbor, close to the shelter of the Isle of Portland (Vindelis), and the distance of Dorchester from any other port, must have rendered Weymouth a most convenient and necessary naval station, during the residence of the Romans in Dorsetshire. The nearest rising grounds on the north-west and north-east of Weymouth are strewed with fragments of Roman buildings, tesseræ, bricks, pottery, and tiles, and Roman copper coins. A large handsome Roman pavement was laid open, and covered up again by King George the Third; and Mr. Medhurst has recently discovered the foundations of several villas, of a Roman temple and of a Roman road. Dr. Buckland supposes these villas to have been occupied by the families of Roman officers or civilians connected with their great military establishment at Dorchester. The most remarkable discoveries made by Mr. Medhurst in 1843, and visited in October last by Dr. Buckland and Mr. Conybeare, were the foundations of a temple on the summit of Jordan Hill, and of a villa, a quarter of a mile distant, in the meadow between this hill and the village of Preston.

"The temple appears to have consisted of a cella twenty-four feet square, surrounded by a peristyle, the walls of which inclosed an area of a hundred and ten feet square. In the earth which occupies this peristyle Mr. Medhurst found more than four sacks of bones, and many horns (chiefly of young bulls), also many Roman coins, fragments of Roman Pottery, cement, &c. Near the centre of the south wall were the foundations of steps, indicating the ascent to the door of entrance; and four feet in advance of this wall are the foundations of four small columns. A layer of cement, which probably supported a pavement that has been removed, occupies the interval between these pillars and the foundation of the south front wall. Within the temple, in the south corner, was a dry well fourteen feet deep, that had been filled in a very curious and unexampled manner. It was daubed all round with a lining or parjeting of clay, in which was set edgewise (like Dutch tiles round a fireplace) a layer of old stone tiles, which, from their peg-holes, appear to have been used or prepared for use on roofs of houses; at the bottom of the well, on a substratum of clay, was a kind of cist formed by two oblong stones, and in this cist two small Roman urns, a broad iron sword, twenty-one inches long, an iron spear-head, an iron knife and steel-yard, two long irons resembling tools used by turners, an iron crook, an iron handle of a bucket, &c., but no bones. Next, above this cist, was a stratum of thick stone tiles, like those which lined the well; and upon it a bed of ashes and charcoal; above these ashes was a double layer of stone tiles arranged in pairs, and between each pair was the skeleton of one bird, with one small Roman coin; above the upper tier of tiles was another bed of ashes. Similar beds of ashes, alternating with double tiers of tiles (each pair of which inclosed the skeleton of one bird and one copper coin), were repeated sixteen times between the top and the bottom of the well; and half-way down was a cist, containing an iron sword and spear-head, and urns like those in the cist at the bottom of the well. The birds were the raven, crow, buzzard, and starling; there were also bones of a hare.

"Dr. Buckland conjectures that this building may have been a temple of Esculapius, which received the votive offerings of the Roman families and invalids who visited Weymouth for sea-bathing and for health; the bones of young bulls found in the peristyle being those of the victims offered in ordinary sacrifice, while the smaller birds, whose bones are found so remarkably arranged in the well, may have been the votive offerings presented by those who received their cure from sea air and sea-bathing, and possibly from the mineral waters of Radipole and Nottington, all in the salubrious vicinity of a temple, which there is such professional reason for supposing to have been dedicated to Esculapius.

"Dr. Buckland then described the remains of a villa in a meadow between Jordan Hill and the village of Preston, and exhibited specimens of tiles, both stone and brick, and various bones and the claw of an eagle, found in the ruins of this villa. In some fields, also, near Radipole, on the north-west of Weymouth, Mr. Medhurst has discovered Roman urns and human bones, and conjectures the spot in which he found them to have been used as a cemetery. The contiguous fields are covered with fragments of Roman bricks, pottery, and copper coins. One gold coin of Constantine, discovered here some time ago, is in the possession of Mr. George Frampton.

"Mr. Duncan expressed his approbation of the supposition that the remains of the large building are those of a temple of Esculapius; but he was unable to account for the pieces of money found with the skeletons of the birds, &c."

At a subsequent meeting of the Ashmolean

Society, Dr. Buckland continued his subject. I have all his MS. notes, which he collected at considerable pains; but I prefer giving his account as it was printed under his own direction at the time. We read:—

"Dr. Buckland next exhibited a part of a human lower jaw, a clavicle, astragalus, and portion of a pelvis, found in the Roman grave by the side of the Roman road, four miles south of Dorchester, on the way to Weymouth, mentioned at the last meeting. The age was indicated by the small size of the bones and jaw, and by the posterior molar or wisdom tooth not being fully formed.

"He also exhibited many fragments of coarse dark-colored Roman pottery, and of Roman mortar (containing pounded bricks), and fragments of the bones of swine and young bulls found in the peristyle, containing in their cavities ashes mixed with earth; but none of the bones found here bore the least marks of fire on them.

"Dr. Buckland then spoke of the absence of

evidence in classic writers as to the disposal which the Romans made of the bones of animals offered in their many and sumptuous sacrifices. The Greeks, he observed, destroyed them with fire. He then quoted Homer's description of the funeral pile of Patroclus: on the top was placed the body, then were killed a number of sheep and oxen, from which having taken the fat and spread it over the corpse, they placed the carcases around the pile, with vessels filled with honey and oil. They next laid on four horses and two favorite dogs, and twelve Trojan captives. In another passage, Achilles says it would be easy to discover the remains of Patroclus, from their place in the centre, whilst the other men and beasts would be found intermixed near the circumference. In a subsequent passage, the bones of Achilles himself also are said to have been distinguished in the same way:

> 'Εν μέσσῃ γὰρ ἔκειτο πυρῇ, τοὶ δ' ἄλλοι ἄνευθεν
> 'Εσχατιῇ καίοντ' ἐπιμίξ, ἵπποι τε καὶ ἄνδρες.

"Dr.Buckland also exhibited the claw of an eagle found by himself in the ruins of the adjacent Roman villa, and, recurring to the conjecture he offered at the last meeting, that these were bones of sacred birds connected with augury, and probably with votive sacrifices to Esculapius (of which we have an example in the cock which Socrates, in his dying moments, commanded to be sacrificed to that deity), and having quoted passages from Horace, in which the raven is mentioned amongst birds that foreboded by the voice, he read, from Littleton's Dictionary, the following passage :—'*Oscines* aves Appius Claudius esse ait, quæ ore canentes faciunt auspicium, ut corvus, cornix, noctua, parra, picus : *Alites*, quæ alis ac volatu, ut buteo, sanqualis.'

"Dr. Buckland then stated that, considering the universal application of birds to the purpose of augury and of sacrifice, by the Greeks and Romans, from the time of the Trojan war to the destruction of the Roman Empire, it was surprising there were so few allusions in classic authors to the ceremonies of sacrificing them, and no reference at all to the disposal made of the bodies either of the wild sacred birds of omen and of sacrifice, or of the bodies of the cocks and hens, which were not only used in their ordinary temples, but were carried about with their armies. Further, we know not whether the same individual fowls, being considered sacred, were kept on through life either in the temples or as travelling appendages to the Legions, and after death buried with religious ceremonies, as appears to have been the case in the temple at Christ Church; or whether the fowls were changed continually and formed the temporary stock of poultry for the table of the priests. Nor have we any reference to interments of ravens, crows, buzzards, and starlings, in a well like that discovered on Jordan hill.

"Dr. Buckland next read an extract from Dugdale's 'England and Wales Delineated,' stating that the late Gustavus Brander, in tracing the ichnography of the priory of Christ Church, Hampshire, which is supposed to have been the previous site of a Roman temple, discovered within the foundations a cavity about two feet square, which had been covered with a stone, cemented with lead into the adjoining pavement, and containing about half a bushel of the bones of birds, such as herons, bitterns, cocks and hens, mostly well preserved.—He then quoted, from a letter of Professor Owen, a list of the birds whose bones he had identified in the small handful Dr. Buckland had submitted to him; which were all that had been preserved out of nearly a bushel found by Mr. Medhurst, in the curious, dry, sepulchral well, within the Roman temple, near Weymouth. These bones were of the raven, crow, buzzard, and starling: in the same well were found the bones of a hare."

Dr. Buckland has subsequently discovered the following examples of funeral honors offered to the raven and the crow, by the Romans and Egyptians.

"Et Romani Corvo dedere exequias, præcedente tibicine duorum Æthiopum humeris elato." [And the Romans performed funeral rites to the Raven —a flute playeı leading the procession, borne aloft on the shoulders of two Ethiopians.] "Circa Myridis paludem Cornicas atque Ibidis sepulchra sumptuoso lapide visebantur."—*Alexander ab Alexandro*, lib. vi. cap. 14. [Around the Marsh of Myris sepulchres of the *Crow* and the *Ibis* made of valuable stone were visited.] It appears, also, from the following passage, that eagles were offered in the more sumptuous imperial sacrifices: "Quod si Imperator rei daret operam divinæ, centeni feriebantur leones, aquilis nec paucioribus." *Cœlius Rhodig.* lib. xx. cap. 6.

One more note to show that the modern Greeks use sacrifice of birds to this day. The letter is from a gentleman whose signature I cannot read.

"Dear Dr. Buckland,—As you are interested in the registering of all that relates to the sacrifice of birds, I send you a passage I have just met with in 'Ainsworth's Travels and Researches in Asia Minor,' vol. i. p. 131. 'One morning, in the midst of this scarcity, we were surprised to find in front of the house a cock newly killed, but not eaten. Upon inquiring, the Greeks said they had killed it in order to propitiate the genius of the mines, and a sacrifice must not be eaten.' This is a remnant of a very old superstition; for cocks were sacrificed to Pluto by the ancients in a similar manner. Mr. Ainsworth gives no authority for this statement."

I must now return to the Gamekeeper's Museum, from which subject I fear I have sadly digressed.

Tastefully arranged in rows were the tails of pole-cats, stoats, and weasels. Imagine the strong and powerful smell of these little animals to be quite enough to prevent the keeper bringing them home bodily, as a larger trophy. The mischief done by them to game and rabbits is very great, but they are easily taken, and many fall victims to the gun; for when once seen by the gunner, whether in hedgerow, rabbit-burrow, or thicket, he need only sit still within shot, and by imitating the cry or squeak of a young rabbit, bring the intruder almost up to the muzzle of his gun. Wonderful is it how these little animals kill hares and rabbits, but it seems that when pursued by them a rabbit loses its senses, becomes fascinated as it were, and through its simple folly falls an easy prey.

Sucking the blood, and eating merely a portion of the neck and head, and despising the rest of the spoil, the stoat or weasel requires more than one victim for a meal, and hence is doubly destructive. Though so destructive, they are easily trapped, owing to the extraordinary manner in which they stick to a certain line of country.

The best trap is a little steel trap covered over with a couple of bricks, and four more placed as sides, so as to form as it were the entrance to a drain or a sort of gallery. This judiciously placed will not fail to catch any passing stoat or weasel, without any bait whatever.

A friend, a good observer of nature, writes to me as follows :—

"I will mention one curious instance of the voracity of a stoat. Last March year, at Castle Forbes, my servant was throwing the casting-net in to catch me some live bait near one of the coverts, when he saw a rabbit start out, evidently ill at ease; it paused, went on, stopped again, till it got two hundred yards from the covert. Then came out a stoat, hunting on his track, and the rabbit actually stopped, as it were petrified, till the stoat jumped on his back. My servant ran up, tried to kill the stoat with a stick, but it got away back to the covert; however, he secured the rabbit, which was slightly bitten on the neck and disabled.

"About half an hour afterwards, within twenty yards of the same place, out came another rabbit in precisely the same way—the stoat following. Again my servant attacked it—the stoat ran between his legs and got into the covert; he secured the rabbit as before. Shortly afterwards, I came up to see if my servant had caught me any live bait, and he was in the act of relating what he had seen, when out came from the same covert a third rabbit, in precisely a similar manner, and hunted by the stoat. My gun was on my arm (as it always is in that wild district,) but I let the stoat kill the rabbit, and then I shot him; thus bagging three rabbits, and killing the stoat into the bargain."

The same gentleman witnessed an extraordinary encounter between a stoat and a cock-pheasant in an open field. The pheasant ran a few yards, then awaited the attack of his enemy; the stoat followed and sprang upon him, and was received with a more vigorous welcome from his spurs.—Again the pheasant ran, and the stoat pursued—another encounter, and similar reception. This was repeated half a dozen times across the field, without any apparent advantage to either side—but our friend had them well within range, and put an end to the combat by shooting Mr. Stoat, and by sparing Mr. Pheasant.

I have heard that the poachers sometimes put down food for the pheasants in an open place in the covert, and when these birds have arrived, they let loose a fighting-cock or two among them armed with sharp steel spurs. The pheasants have no chance against these horrible weapons, and so often get killed by the fighting-cocks, and bagged by the poachers. All this can be done with very little noise, so as not to attract the gamekeepers.

It is often a difficult matter to know which of a

lot of birds, pheasants or partridges, hanging in a larder, ought to be cooked first. Mr. Coulson, of Clifton, Bristol, has shown me how to put a date upon each bird, without using pen, ink, or pencil, and it is a very simple but useful plan. When the birds are brought in after shooting, hold up each before you with his breast facing you—then begin to count his toes from your right towards your left, after the manner that children in the nursery play the game of "Whose little pigs are these?" Let the claws indicate the days of the week; if the bird was shot on a Monday, pull the claw off the first toe you count—if on a Thursday, the claw from the fourth toe, and so on. When the birds are subsequently examined each will bear a mark to show immediately on what day of the week he was killed. This plan may be known to many, but still I give it for the benefit of those who have never heard of it before.

In our keeper's museum we observed two rows of dead hedgehogs, numbering in all one hundred and thirty. These had been executed for the crime of egg-eating. For many years I had upheld the cause of the hedgehog; but I fear the verdict of "Guilty upon the above plea" must be found against him. A hedgehog will eat almost anything. He will eat dead birds, roots, beetles, and slugs. I have often seen the heaps of cow-dung by the sides of coverts turned over. The hedgehog has been at work looking for the beetles and worms which are found under them; but I am sorry to say an egg is to him a most savory feast.

I lately obtained a hedgehog, and gave him some bread and milk and an unbroken egg. The egg was the first course of his dinner he attacked—upon examining the basket a few hours afterwards we found the egg entirely gone—not one little bit of the shell was left; he had eaten up every bit of it. This, we think, will fully account for the keeper's not finding any broken shells in the pheasant's nest.

A hedgehog will not only eat eggs, but he will also eat flesh; and if he can't find anything dead to eat, he will kill some on his own account.

Now, a hedgehog is not a very powerful animal; he therefore of necessity attacks something weaker than himself, such as young partridges and leverets. We will again put a witness of great experience in the box. He was out by a covert side with his keeper, when the terrier dog set at something in the ditch; he heard a terrible squeaking at the same time. On turning over the long grass and brambles that overhung the ditch, he found a leveret, about the size of a rat, in the jaws of a hedgehog, palpitating in the struggle of death. The sharp teeth of the hedgehog are adapted equally for catching a small animal, such as a leveret, or for munching up with ease the hard and horny cases of beetles.

Mr. Francis Francis, in an admirable letter in *The Field*, gives the following evidence as to this point:

"Several instances have been forwarded to *The Field*, and some few have been related and sent to me privately, wherein hedgehogs have been proved to have broken into hen-roosts, and abstracted young chickens, ducks, turkeys, and also to have taken young pheasants, and other kinds of game; and these instances have not been one or two, but more than a dozen in number, and from all parts of the country. One instance of a hedgehog carrying off two partridges, quite two thirds grown, came under my knowledge; and well-authenticated instances of their voracity are so numerous, that I cannot doubt that to an extent they may be looked upon as vermin and destructive to game. That their prevailing characteristics are carnivorous is easily discernable by the teeth. Upon two occasions those in my possession killed and ate one of their number; and on the second occasion the individual killed and devoured was fully two-thirds grown, and as large as any one of the family, excepting the old one. In one night there was nothing left of him but the bare skin and prickles. Now, if a hedgehog is capable of devouring another one of his own size, body and bones, will any one pretend to find a reason why he cannot accomplish the same feat upon a partridge or a young pheasant?"

Early in the spring, I tried to buy a hedgehog

in London, but could find one neither in Leadenhall, St. Giles's, nor in other animal markets; they were not to be had for love or money. A few days afterwards, the warm weather came, and I found plenty of hedgehogs for sale. The *probable* reason of this fact is, that the hedgehogs had been concealed in their winter quarters during the cold weather, and that, coming out in the warm, had been caught by the dogs of those curious fellows who poke about the hedges near London for snails, snakes, birds' eggs, &c.—all saleable articles in St. Giles's animal market.

At Castle Forbes, in Ireland, there are many hedgehogs; and there they are caught in a box-trap, made of iron bars, called an "iron witch." During the spring of last year, more than fifty of them were caught by one of these iron witches, the bait being always a rabbit's paunch. Like the cats, the hedgehogs are very fond of valerian, and, with their sharp sense of smell, soon find out a trap baited with it.

A gamekeeper at Ringwood informed me that he trains retrievers to catch hedgehogs for him, and he goes out at night to hunt for them. He has killed upwards of thirty in one night by trampling on them. (N. B.—Nearly all these animals found in the museum have been thus served, and I always find it most difficult to get a perfect skull from the collection.) The keeper informed me that the hedgehog "eats black-bobs—them great beetle things;" that they resort to the same place to feed, night after night; and that they "routs up the cow-dungs for the black-bobs;" and that a stinking bait is best for them. The otters, too, have also regular hunting-places, and they have beaten down paths and tracks across the interspace of land about two miles between the rivers Stour and Avon. When a trap is set for them, it should never be fixed; because if it is loose, it will float in the water when the otter retreats there, and will help to kill him.

I have often heard that hedgehogs are good to eat, and that gipsies are very fond of them, and that they are great proficients in the art of cooking them. I have lately had the good fortune to obtain information on this point from a high authority. In the neighborhood of Oxford I met an old gipsey woman, who, although squalid and dirty, was proud in being able to claim relationship with Black Jemmy, the king of the gipsies. She informed me that there were two ways of cooking a hedgehog, and seemed much surprised at my question whether her tribe ever ate them; as if there could ever exist a doubt. I expressed a wish to know the process, the receipt for which I subjoin in her own words: "You cuts the bristles off 'em with a sharp knife after you kills 'em fust, sir; then you sweals them (Oxfordshire, burns them with straw like a bacon pig,) and makes the rind brown, like a pig's swealings; then you cuts 'em down the back, and spits 'em on a bit of stick, pointed at both ends, and then you roasts 'em with a strong flare."

It appears that hedgehogs are sometimes in season, and sometimes out of season. My informant told me that "they are nicest at Michaelmas time, when they have been eating the crabs which fall from the hedges. Some," she added, "have yellow fat, and some white fat, and we calls 'em mutton and beef hedgehogs; and very nice eating they be, sir, when fat is on 'em."

The other way of cooking hedgehogs is gone out of fashion. The gipsy's grandmother used to cook them in the following manner; but it appears they are best roasted. The exploded fashion is to temper up a bit of common clay, and then cover up the hedgehog, bristles and all, in it,—like an apple in paste, when an apple-dumpling is contemplated,—then hedgehog, clay and all, is to be placed in a hole in the ground and a fire lighted over it; when the clay is found to be burning red, the hedgehog is done and must be taken out of the hole; the clay-crust of the pie being opened, the hedgehog's bristles are found sticking to it, and the savory dinner is ready.

The fashion of eating hedgehogs was not, in former days, confined to gipsies. There was a farmer's family living at Long Compton, near Oxford, who were supplied with hedgehogs by our informant's grandmother; this family used also to breed them, keep and fatten several litters, "and," said the gipsy, "they used to eat up every litter they bred, dressing 'em just when they wanted 'em, like they did the fowls." Sometimes a nest of young hedgehogs is found by the gipsies; if they are too small for eating, they are preserved till fit for use, or, as it is called in Oxfordshire, "flitted;" that is, a string is tied to the hind leg, and the doomed animal is allowed to wander about the length of his tether, picking up what he can get; under this system, if well fed, he will fatten wonderfully.

It has long been a disputed point as to whether hedgehogs will eat the common harmless snake or not. There is no reason why they should not, as their teeth are sufficiently sharp and pointed both to catch the snake and munch him afterwards. Again, it might be argued that the snake would be too swift for his bristled enemy, and be able to escape by flight. This is not the case; for, in the first place, the crafty hedgehog might come upon the snake when basking in the sun; and even supposing the alarm was given, and a pursuit took place, the hedgehog would have the best of the race. It is quite surprising how fast a hedgehog can run, if he likes; all his bristles lie quite smooth on his back, his little legs generally coiled up tight in the centre of his body, and he hustles along at an amazing pace; and, in a fair race in the open, with a fair start (a difficult thing it is to start a hedgehog for a race, as I have tried, the brute will persist in rolling himself up and *not* starting), we would back the hedgehog against the largest snake in England. I therefore determined to try the experiment, whether a hedgehog would really eat a snake or not. I caught a snake near Harrow, and bought a hedgehog in St. Giles's. For several mornings I placed them together on the grass, but they took no notice of each other. The snake never went *towards* the hedgehog, if he could help it; if he had a choice, he went in the contrary direction; but, whether through shyness at my presence, or whether because he was not hungry, the hedgehog never attacked him. At last, one evening, I shut them both up in a box together. The next morning, the lid was opened, and the murder was discovered.

The hedgehog had, during the night, attacked and eaten half the snake, beginning at the tail. He had not left a single bit of the lower half of the body, and instead of the fine active snake I had put in in the evening, I found only half a snake was left; just as if he had been cut in two with a knife, and the lower half taken away. As the hedgehog had begun his meal, I thought I would allow him to finish it, which he did in a very few hours. We must at the same time be guarded against the conclusion that we are dealing with a great serpent slayer; and a little reflection upon the habits of the individuals in question will tend much to set us right upon this point.

The snake loves the sun, the hedgehog is a nocturnal animal; he is very seldom or ever seen out looking for his food before the sun is down; he then comes out of his den, and begins hunting for beetles and worms—particularly the big lob-worms that come out of their holes to enjoy the dew of the evening, as every fisherman well knows. If he came out in daytime, he would find no worms, and but very few beetles, even if he were to look for them, which, by the bye, he would have some difficulty in doing, as his eyes are of a dark color, and more suited for twilight than sunshine. The snake, on the contrary, comes out only in the heat of the day, to bask in the sun, or hunt in the shade of the long grass for frogs; these two animals are therefore not very likely to come across one another; still such a thing is possible, and does, we believe, occasionally happen—the rencontre taking place between an early hedgehog and a late snake. The hedgehog is, *par excellence*, one of the scavengers of our fields; and although his staunchest supporters may not be able to clear him from the foul stain of occasionally indulging in a pheasant's egg, garbage and animal refuse form his general repast, seasoned with insects. It is very probable, too, that the hedgehog appropriates to himself the wounded game. After a day's covert shooting, many a wounded pheasant, hare, and rabbit, mortally struck, escapes from the dogs and the beaters, to retire into some quiet hiding-place to die; the hedgehog finds them out, and, if half dead, performs an act of mercy in putting them out of their misery; if he finds them quite dead, he is saved his trouble, and makes a good supper, devouring that which would shortly become putrid, and do harm to the remaining living inhabitants of the covert.

There is a slight difference in the color of the sexes of the hedgehog, as well as in the size; the male being longer and darker. Both are tainted with a decided musky smell, and the exuviæ of the hedgehog can always be distinguished by this peculiar odor. Some years ago, when routing about an old ruined castle, situated in the centre of a pine forest in Germany, I discovered among the ruins a fine specimen, which illustrated the above statement. Many a naturalist has since been puzzled with our specimen.

The baby hedgehogs are the funniest little things possible; they are born covered with tiny spines, which are quite soft almost like hair. If touched, their natural instinct prompts them to curl up. This they cannot do, as the beautiful yet complicated set of muscles whereby they are enabled to perform this operation are not developed till the spines acquire some degree of hardness.

A female hedgehog was bought from a boy a few weeks ago, in the neighborhood of Oxford, and placed in a basket: in a short time, four baby hedgehogs made their appearance; but their cruel mamma devoured all her progency, leaving not a bone or a bristle. It appears, however, that her meal disagreed with her, for she shortly afterwards died herself, her children not agreeing with her parental stomach.

The bristle of the hedgehog, if cut across, will be found to be quite hollow; the walls are formed of a hard horny substance, and the interior is filled with a sort of pitch—as pith lies in a stem of elder. It is, in fact, nothing more than a magnified human hair; and a human hair, under the microscope, looks very much like a hedgehog's bristle when viewed with the naked eye. We know only of two uses to which the hedgehog's spiny coat is put by his enemy, man. Coachmen will sometimes tie a hedgehog's skin on to the pole of the carriage, to prevent a shirking horse from leaning against it; and we have seen the single spines used by the German professors of anatomy, to pin out dissections of nerves and muscles. These preparations are often placed in bottles containing a corrosive fluid; and were the pins used made of metal, they would shortly corrode, and spoil the preparation. In the hedgehog's bristles they have ingeniously found a natural pin, which serves the purpose admirably. I have often wondered they have never been made use of by English ladies. The North American squaws ornament baskets, mocassins, &c., with porcupines' bristles, and hedgehogs' bristles are not very unlike them.

I cannot conclude this article without mentioning, in short, a few hints given to me by a friend, who has so kindly supplied me with much of the matter I have embodied in the previous pages, relative to the science of game-keeping, in which he is a great practical authority.

The unthinking, unobserving gamekeeper shoulders his gun, and walks round his coverts and woodlands, disturbing everything, and tires out his legs instead of using his brain; in fact, he defeats his own object, and gives himself much trouble into the bargain. To know if all is quiet in a covert, the observant sportsman should quietly steal into it, and station himself where he can have as extensive a view as possible. Having chosen his point, he should remain there as still as possible, with his eyes wide open and ears all attention. He should go early in the morning, or else in the evening, when the animals are all out at feed; and he should be particularly careful not to let them get scent of him. If he sees the game all out, quietly feeding,—the hares and rabbits unalarmed, the pheasants picking about at their ease, the wood-pigeons flying lazily to and fro among the trees,—he may conclude there is no enemy about, either vermin or poacher. A good telescope, or, what is better still, a good pair of double race or opera glasses, are exceedingly useful when on the look-out. The only drawback to their use is, that (as I have understood) statements of facts and deeds observed through a telescope will not be received by county magistrates as evidence. If the watcher sees no hares or rabbits, and no game out of the covert feeding, he may be quite sure that there is something about to disturb them; particularly if he sees (as he probably will) a starled hare or rabbit, fleeing from danger, come cantering along, stopping every now and then to listen, he may be certain there is something behind in pursuit, and should take his measures accordingly.

He should look at all footmarks, both of man and beast; and if he sees the print of a boot, he should know who has made that print, and whether the owner has any business going along there. He should watch the actions and countenances of the people laboring in the fields, and take mental notes at every gap in the hedgerow, every footstile in the path, and every gate in the roadway. In fact, the science of game-keeping requires a good head, a sharp eye, an accurate observation of the commonest things, and a ready application of what has been observed. It is a talent that can never be acquired, unless it is *in the man* at starting. Observation, and accurate observation too, is the foundation of success.

I will give an instance. A gamekeeper is sound asleep in his bed. He suddenly wakes, and *fancies* he has heard a gun in the coverts; but, being still half-asleep, he is not sure, and even if he did hear the report distinctly, how is he to know in what part of the covert it was fired? In an instant, he should (without attempting to dress) rush to the house-door, and opening it, look out sharply towards the moon, or, if there is no moon, towards that part of the sky where there is most light. The chances are that, in a minute or two, a frightened bird (most probably it will be a wood-pigeon) will pass between him and the light; he will note from what direction it is flying, and will start immediately in that direction to look for the poacher. The bird would not be flying in that course, if he had not been startled from his roosting-place by some intruder.

Observation, too, is most necessary for those who delight in fox-hunting; and people often get lost in strange countries for want of it. A master of fox-hounds—a specimen of the olden stamp, full of years, good health, and sound common sense—started one morning from his house to meet the hounds that had gone on to the coverts some time previously. He was accompanied by a gentleman, in whom the power of observation was by no means developed. The squire soon struck off the main road, through a bridle-gate; then down a hedgerow; through a gap; then up a wet and muddy lane; then through a farmyard; then from one corner of a ploughed field to the other corner, where there was no exit without jumping a ditch; and so on, through most out-of-the-way places, but yet always on a straight line.

At last his friend ventured to remark, "Are you sure you are right, squire?" "Right! Of course I am right," was the answer. And on they went again, straight across the country. Again the friend said, "It's very odd; I don't see any road at all. Where are we going?" "To the meet, to be sure. The hounds have gone along here." "The hounds gone along here! Why how can you know that?" "*I can smell them*," growled the squire, "quite plain; can't *you?* They have only been gone on an hour, and that's nothing." The visitor sniffed and sniffed, but could perceive nothing. "I don't suppose *you* could," remarked the squire; "it's only the master that can smell his own hounds: and you see I have been right all along, following my nose, for there they are by the covert's side."

In the evening, after dinner, the visitor was telling a long story of the wonderful power of the master's olfactory organs,—how that he followed the hounds to covert *by their smell*, and did not make a mistake the whole way. The story was passed round the table, and a discussion began among the younger hands; the old stagers held their peace, or narrated still more wonderful stories; till at last the squire let the "cat out of the bag," and explained that he had not smelt the hounds at all, but *only followed their footsteps in the mud*, from the highway where they turned off towards the meet by the covert side. Shouts of laughter ensued at the discomfited visitor, now that his wonderful story had exploded; and he packed up his bag next morning, thinking what a goose he had been not to have better trained his powers of observation of common things.

IN MEMORIAM.

IN MEMORIAM.

Among my father's papers were a few copies of the following beautiful verses. They were written, as appears by the date upon them, December 1st, 1820, about one year after his election to the professorial chair of geology at Ox ford.

Dr. Buckland, at that time in the fullest vigor of mind and body, stood forth as the champion of geology, then a new science in the University. He created a taste for its study in the minds of young and old. Heads of houses and undergraduates flocked to his lectures, and alike became admirers of the Professor and his pursuits. Alas! how few of his class are now alive; the teacher and the students alike have passed from among us.

One of the class who assembled in the Clarendon Museum, December 1, 1820, was so struck with the lectures that he had heard, that he composed the following lines, which he has kindly allowed me to reproduce and publish in this little book.

I here beg to thank him for his kind permission, and to hope that he may be long spared to enjoy good health and the high position which he has won by his great talents and exemplary character. The lapse of forty years has added meaning and force to his lines; and some of them may almost be said to be prophetic.

Mourn, Ammonites, mourn o'er his funeral urn,
Whose neck ye must grace no more;
Gneiss, Granite, and Slate! he settled your date,
And his ye must now deplore.

Weep, Caverns, weep! with infiltering drip,
Your recesses he'll cease to explore;

For mineral veins, and organic remains,
No Stratum again will he bore.

Oh! his Wit shone like Crystal! his knowledge profound
From Gravel to Granite descended;
No Trap could deceive him, no Slip could confound,
Nor specimen true, or pretended:
He knew the birth-rock of each pebble so round,
And how far its tour had extended.

His eloquence roll'd like the Deluge retiring,
Which Mastodon carcases floated;
To a subject obscure he gave charms so inspiring,
Young and old on Geology doted.
He stood forth like an Outlier,—his hearers admiring,
In pencil each anecdote noted.

Where shall we our great Professor inter,
That in peace may rest his bones?
If we hew him a rocky sepulchre,
He'll rise and break the stones;

And examine each Stratum that lies around,
For he's quite in his element under ground.

If with Mattock and Spade his body we lay,
In the common Alluvial soil,
He'll start up and snatch those tools away
Of his own Geological toil.
In a Stratum so young the Professor disdains
That embedded should be his Organic Remains.

Then exposed to the drip of some case-hard'ning spring,
His carcase let Stalactite cover,
And to Oxford the petrified sage let us bring,
When he is incrusted all over;
There, 'mid Mammoths and Crocodiles, high on a shelf,
Let him stand as a Monument raised to himself.

1*st December*, 1820.

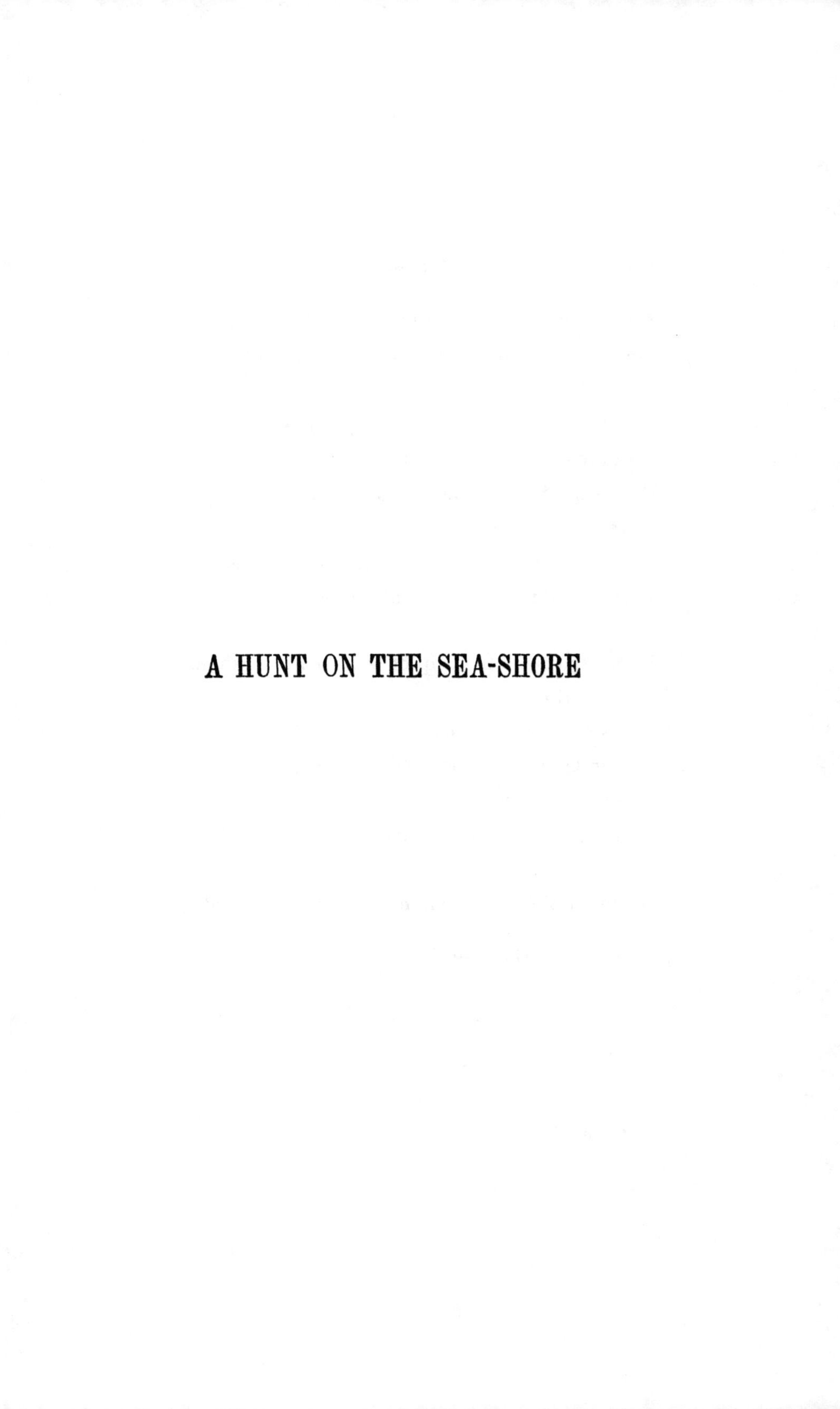

A HUNT ON THE SEA-SHORE

A HUNT ON THE SEA-SHORE.

"A TELEGRAM for you, sir." Very well. "Wanted at Brighton immediately, to see an invalid." My portmanteau is soon packed, and as I roll over London Bridge, looking out for the White Horse that is proverbially always to be seen on the bridge, I think how soon I shall change the view of the muddy river for that of the open sea, and the rattle of cabs for the rattle of the pebbles on the beach.

Comfortably seated in the carriage, I remark to my companion, as we travel over the tops of the houses close to the railway-station, that it is generally possible to know upon what geological formation a town stands, by observing the substance with which the walls of the houses are built. Thus, Bath is built principally of the stone found so plentifully close by; Oxford, of oolite, of which stone there are beds in abundance close at hand; London is built principally of bricks made from the London clay, upon which, for the most part, it stands. The time rolls quickly by, and we are at Brighton. Now let us apply this theory. Chalk, chalk everywhere. People cannot build houses of chalk! I wonder (for it is my first visit) what the Brighton folks use for this purpose. I am not kept long in suspense; one of the first houses that meets the eye, in going out of the station, is built, not of chalk, but of something that comes out of the chalk—viz. flint. We examine the walls of the house; we find there flints of two kinds—those worked by the hand of man, and those prepared by the sea; the former are, for the most part, nicely squared, the latter as nicely rounded, and, strange to say, all about the same size, as though picked out by a careful sorter. Who this sorter was, we shall see presently.

But how do the Brighton people manage for

pavement? They have no stone quarries near, so they use what they have at hand—viz. clay. The bricks are of a bright red color, showing that they contain much iron. In places, however, where there is a great deal of traffic, as on the Marine Parade, we observe that they have placed down layers of small flag-stones and bricks alternately—parallel, however, with the direction of the road—presenting a magnified view of the structure of an Indian elephant's molar tooth; and, in consequence of this arrangement of bricks and stones, a curious thing has happened: the softer material, the brick, has worn away into hollows, leaving the harder, the stone, projecting; thus puddles are formed at the crossings, to the discomfort of the passengers and the advantage of the crossing-sweepers. It would have been better if the crossings had been made either entirely of brick, or entirely of stone, as the wear and tear would have been equal in all parts, and no holes formed.

Brighton seems to be the Paradise of invalids, and, according to the old maxim that demand creates supply, we find that chemists' shops have sprung up in all directions. People who have over-eaten themselves, evidently come down here to look for good digestion, or repair that which they have already damaged, for we see advertisements in the windows, of medicines done up in a palatable form, such as "Magnum bonum jujube," "Digestive dinner tablets," &c. People whose lungs are unsound also come here, as is evident from the many kinds of respirators we see exposed for sale, from the five-shilling one of cork and cotton, to the higher priced article of gold and silver.

That the visitors to Brighton bathe in the sea is also pretty clear, from the army of bathing-machines we see drawn up along the shore, above ordinary high-water mark (it was winter when I made these observations); yet the proprietors do not feel quite certain that their property is safe from storms even here, and accordingly they have not placed all their machines with the doors facing the sea: the front rank are turned with their sides towards it, and thus act as a barrier, should the pebbles be washed away from under the wheels of their comrades in the hindermost ranks, and they should spontaneously march off down the inclined beach into the deep water, according to their summer wont. While looking at the machines, I was informed by my companion that the English have not been a sea-bathing nation such a very long time, and that, therefore, bathing-machines are a comparatively modern invention. It is exactly one hundred and ten years ago that a physician, named Russell, wrote a book upon the advantages of washing the body in sea-water—an idea which had not previously entered into the brains of our forefathers. Up to that time, to use the words of my learned informant and friend, Mr. Roberts, of Dover, "the sea was judged to have been designed for commerce, and sea-side towns for the residence of merchants and fishermen. At no previous period had there been sea-side visitors. Why should they go to the sea-coast, when no motive could be stated,—at a time, too, when Northampton's healthy climate was attributed to its distance from the *noxious fumes of the sea?* There were certainly watering places; but these were towns where mineral waters existed, such as Bath, Cheltenham, Harrowgate," &c. Dr. Russell's brother doctors took up the cry; sea-bathing suddenly became the fashion; Dr. Russell was obliged to come to reside at Brighton; and the fishing villages in various parts of the kingdom became inundated with visitors. Brighton, being the point where the sea could be most easily reached from London, was soon found out, and taken possession of by a colony of citizens, anxious to follow the fashion and recruit their health at the same time. Besides Brighton, many other watering-places have started into existence, and the sea is now found efficacious for nearly all ailments, whether of mind or body, and it often effects a cure when nothing else will; an annual migration, like that of anadromous fishes, of thousands of persons now takes place to those very shores which their grandfathers regarded with a species of horror.

In most books, Brighton is stated to be forty

odd miles from London. This we believe not to be strictly correct; but it is *made under* fifty miles from London, because, as we have heard the tale, in former times, the King was not allowed to go more than fifty miles from London without a minister. Now, the sovereign who was so fond of Brighton did not want to be bored with a minister at his elbow; and therefore Brighton was put down as being under the proscribed distance, and the Pavilion, &c., started into existence.

The general complaint of a sea-side place is, that there is nothing to do; and we see ladies and gentlemen basking in the sun, and lounging on the piers, reading shilling novels and tales of love and murder, surrounded on every side by objects which, if they would only use their eyes, would afford endless amusement and instruction.

I lost no time, the morning after my arrival (business, of course, being first àttended to,) in getting down to the sea-side, and listening once more to the ocean's solemn and never-ceasing song of praise to the great Creator. To the observer, the composition of a sea-beach will tell much. From the Brighton beach we learn that the bottom of the sea, off the land whereon we now stand, is principally chalk; for it is composed almost entirely of flints of every possible form and shape, but all of them more or less rolled, and their edges worn off by the action of the water—for Nature never turns her work out square, always round. As the stonemason picks out the flints from the chalk-pit, pickaxe in hand, so does the sea, by the force of her waves, extract from their chalky bed the flints; but she does more than extract them—she, so to say, manufactures the raw material, and not only manufactures, but even sorts her work, when finished, ready for the use of man. Upon a careful examination of the pebbles on the beach, we find that they are sorted in a very regular gradation. When a wave runs in, it carries with it a certain number of stones, at the same time driving a certain quantity along the bottom. We find that the largest always go the farthest, the smallest stop first. Now, let us apply this. The portion of the beach nearest the shore is composed of the largest stones, ready sorted to the hand of the wall-maker; the intermediate sizes are farthest out, and are sifted by the mortar-mixer to mix up with his lime. Farther out still, and under the water, are the smallest stones of all, namely, the sand. Of this the shrimper takes advantage; for, at low tide, he may be seen, up to his hips, pushing before him his shrimp-net, and disturbing the little denizens of the sand, the shrimps and prawns—condemned shortly to enter water at a considerably higher temperature than they have been accustomed to.

Among the shingle on the Brighton beach, I found many interesting things, which most people would call rubbish. They are of no use in themselves; but imagine the beach to become suddenly fossil, and how interesting would all these bits of rubbish then become, as proving the existence of a highly civilized people who once inhabited these shores. We find, mixed up with fish-bones, sea-weed, &c., various kinds of the jetsam and flotsam of the ocean. Rolled bits of ginger-beer bottles, made of clay; of wine bottles, tobacco-pipes, &c., made of glass, proving the luxury of these ancient people; of old shoes, of knives, of hard slag (whence we might infer the existence of gas-works,) of combs, hair-brushes, &c.; and, what is rather curious, I found a ball, as large as a good-sized turnip, and quite as round, composed entirely of human hair. The only way I can explain the formation of this ball is, that the hair taken from the combs of the inhabitants of Brighton is thrown away, floats down the drains into the sea, and there becomes regularly felted and matted together by the waves perpetually rolling it up and down on the hard shingle. I at first wondered how the harder and heavier articles of human manufacture, such as bottles, &c., became spread in such abundance along the shore, for, they would not float down the drains, like the hair. The problem was solved when I discovered that the rubbish carts of the town, shot their loads over the cliff at the extreme east end of Brighton; the authorities hoping thereby to prevent the cliff falling further than it has fallen already. This then is, I believe, the main source of most of the water-worn human manufactures found spread out, for some three or four miles to the eastward of the town.

The deposit of the river Thames is, likewise, most interesting, and a chapter might be written on the various articles found mixed with the bones of turtle, snakes, and sharks, and even tropical fruits, in the Isle of Sheppy, at the mouth of the Thames. By means of these, an observant explorer landing on the spot from his discovery-ship, even if he knew not of the existence of London, might come to the conclusion that the river which brought the deposits of human manufacture to the shore he was exploring, ran through a large city, inhabited by a highly civilized people, far advanced in the arts. I believe many curiosities of ancient manufacture are found at the bottom of the river in the neighborhood of London Bridge. May not the cause of their getting there be, that they became lost, and thrown over into the water, at the time that the bridge, upon which dwelling-houses were built, once spanned the river at or near this place?

The sea at Brighton not only rolls in flints from the chalk, and sundries from the rubbish-hole, it also washes up lumps of chalk, from the size of a small apple to that of a child's head. These lumps will be found, for the most part, bored through and through with holes, as if an industrious cheesemonger had been working away at them with his cheese-taster. If the inside of these holes be examined, it will be seen marked all over with minute scratches or indentations; it is, as a friend has aptly expressed it, "rifled in the barrel." These perforations and riflings are not, however, the work of man, but of a delicate little shell-fish, called a Pholas. We showed one of these to a fisherman, wishing to get the local name. "Them things, sir? Why, them things be *pitticks;* but I don't expect that's the right name for 'em." There is a large flat surface of chalk to be seen at low tide, to the east of Brighton, perfectly riddled with holes made by these creatures. If an empty pittick's hole be examined, it will be found covered with markings, as though some instrument or some acid had been at work. Now, this is the very point which has set those interested in the subject at loggerheads; but, while they are arguing and disputing, the Pholas quietly goes on working away, making a comfortable hole for himself. On the whole, however, the jury seem inclined to bring the Pholas in guilty of mechanical, not chemical violence, inasmuch as there is no acid found, or other solvent known, that will act equally well upon wood, limestone, hard and soft clays, sandstone, and even, in one case, upon wax; it is also highly improbable that the animal can secrete a solvent for each and every substance in which he may feel desirous to hide his head.

In the "Annals of Natural History," Mr. Albany Hancock says that the excavating instrument of Pholas is formed of the anterior portion of the animal, in the surface of which are embedded siliceous or hard flinty particles; these particles give the skin (which covers the shell) much the appearance of sand-paper. The whole forms a rubbing surface, which, being applied closely to the bottom of the cavity, by the adhesion of the foot, enables the animal to rub down, and so penetrate shell, chalk, wood, &c. Now, Professor Forbes looked for these siliceous particles, and found none; neither the microscope nor the contents of the laboratory bottles would prove their existence. He therefore puts M. Necker in the witness-box, who states, that "the composition of the shell is the same as arragonite, and it is plain that it can act mechanically on the hardest limestone."

If the reader will examine the outside of a Pholas shell, he will find it studded in regular rows, with little projections, which are copied *exactly* on the rough surface of a farrier's file; he will also see models of them on the inside of the common nut-crackers. It is, therefore, most probable that, when the shell wishes to embed itself, it fixes itself firmly by its foot—which acts like the leather sucker used to lift stones by school-boys—and then, by working itself from side to side, presenting the edges of its shell to the substance to be bored, it gradually cuts away a cavity for itself somewhat in the same way as the instrument used by the artesian well-makers cuts through the soft clay. Now, what becomes of the dust, or rather mud, which must naturally result from the boring of the shell? Unless he had some apparatus to eject it, it would accumulate in the hole, and considerably inconvenience the operator; accordingly, we find that he has a very pretty contrivance for ejecting the same. If the reader happens to be in a neighborhood where the Pholas is left dry by the retiring tide, he will have an opportunity of judging for himself; for, as the shells perceive his footstep, they will eject, to a considerable distance, a spout of water, which is generally clear, but which, if they have been at work lately, will be colored with mud, the result of their borings.

Dr. Buckland wrote to the late lamented Professor Forbes on this subject, and the following is part of his reply:—" Pholas, Xyloplaga, and Teredo are mechanical borers, and, as far as I know, do not bore much below low-water mark—very generally above (Pholas especially;) Xyloplaga and Teredo are confined to wood. Pholas bores indifferently in limestone, slate, and wood; and in Mr. Cumming's collection there is a group of Pholades living in wax."

Having mentioned this subject in *The Field*, a gentleman, who signs himself "Astur," interested in the matter, was kind enough to write the following. He was fortunate enough to see the Pholas actually at work, and his observations are most valuable.

"Having seen in the first of Mr. Buckland's interesting communications, entitled 'Sea-side Gatherings,' that it was still a 'disputed point as to how these very delicate creatures (the Pholades) make holes in this very hard stone;' and having had the good fortune to witness how they performed this apparently impossible operation, I feel great pleasure in laying it before your readers, and corroborating Mr. Buckland's hypothesis, that it is by a mechanical process. The mode by which the Pholas bores its habitation, even in the hardest rock, has long been a subject of ingenious speculation among naturalists. None of the theories advanced, however, have been accepted as fully satisfactory: and no one previous to myself has had, I believe, the satisfaction of seeing the animal in the operation. The aquarium, one might have supposed, would have afforded an easy means of solving the problem. But here we meet a fresh difficulty, from the supposed fact that the Pholas, being dragged from its rocky cave, would not take the trouble to excavate another. This theory appeared to me most unnatural and improbable; and my opinion seemed to be further borne out by the fact, that old logs of wood thrown up by the waves were frequently covered with holes of different sizes and depths, from one-eighth of an inch to the entire thickness of the log. These holes appeared to me evidently made by Pholades, and to have been abandoned, either through caprice or necessity. The latter we may suppose to be frequently the case, when the log, resting quietly at the bottom of the sea, is pitched upon by a houseless Pholas, and operations commenced, when the ruthless storm drifts the log on shore, and hurls the Pholas once more at the mercy of the winds and waves. .Here is another case in which the Pholas may require a fresh habitation, for the procuring of which we should suppose Nature had given it the means :—it is this—many of the logs, &c. tenanted by Pholådes are so thin, that they would serve the growing mollusk but a short time, and, moreover, wood and other substances used by then being of a perishable nature, must render a fresh abode frequently necessary. Having, therefore. procured several of these mollusks in pieces of timber, I extracted one, and placed it loose in my aquarium, in the vague hope that it would perforate some sandstone on which I had placed it. It possessed the powers of locomotion, but made no attempt to bore. I then cut a piece of wood from the timber in which it had been found, and placed the Pholas in a hole a little more than an inch deep. Its shell being about two inches long, this arrangement left about an inch and three-quarters exposed. After a short time, the animal attached its foot to the bottom of the hole, and commenced swaying itself from side to side, until the hole was sufficiently deep to allow it to proceed in the following manner. It inflated itself with water, apparently to its fullest extent, raising its shell upwards from the hole; then, holding by its muscular foot, it drew its shell gradually downwards. This would have produced a perpendicular and very inefficient action but for a wise provision of Nature. The edges of the valves are not joined together, but are connected by a membrane; and instead of being joined at the hinge, like ordinary bivalves, they possess an extra plate, attached to each valve of the shell, which is necessary for the following part of the operation. In the action of boring, this mollusk, having expanded itself with water, draws down its shell within the hole, gradually closing the anterior edges until they almost touch. It then raises its shell upwards, gradually opening the lower anterior edges and closing the upper, thus boring both upwards and downwards.

The spines (points) on the shells are placed in rows, like the teeth of a saw; those toward the lower part being sharp and pointed, while those above, being useless, are not renewed. So far for the operation of boring; but how to account for the holes fitting the shape of the animal inhabiting them? To this I fearlessly answer, that this is only the case when the Pholas is found in the rock which it entered when very small. This mollusk evidently bores merely to protect its fragile shell, and not from any love of boring; and in this opinion I am borne out by my own specimens. The young Pholas having found a substance suitable for a habitation, ceases to bore immediately that it has buried its shell below the surface of the rock, &c. It remains quiescent until its increased growth requires a renewal of its labors. It thus continues working deeper and deeper; and should the substance fall or decay, it has no alternative but to bore through, and seek some fresh spot, where it may find a more secure retreat. ASTUR.

Oct. 1, 1859."

Both the north and south piers, or "horns," at Folkstone are formed of gigantic blocks of stone piled closely against one another in a slanting direction, and not cemented together. When a great wave comes thundering in, it does not hit these breakwaters nearly so hard as if they were quite solid, but, on the contrary, gets broken up, and runs through the interstices of the stone; the consequence is, that numerous miniature caves are formed, which are full of animal life. We see the surface of the rocks bored; and these borings contain Pholades, which the fishermen here call "clams." The Folkstone sandstone is so hard that it is with difficulty broken with a hammer; and it is most marvelous how these very fragile and paper-like shells have managed to bury themselves in it. The actual shells crumble under the finger, but yet they have the power of boring clean-cut holes in a substance which a crow-bar will hardly touch.

My friend Mr. Woodward, of the British Museum, very kindly showed me some exceedingly fine specimens of fossil Teredo; one, in particular from the Upper Greensand at Blackdown, presented by my father to the Museum; also a very fine specimen, beautifully polished, of Teredo and Teredo's holes, in a bit of fossil wood from Kew, which belonged to the late eminent botanist, Mr. Lambert. He also pointed out a specimen of the sandstone from the London clay formation, which is dredged up in vast quantities off the town of Harwich, and which is used to make Roman cement. This also contained many Pholades. In this formation the fossil bones of elephants and other large animals are found; the Pholas had attacked these also, but could not bore more than than a certain depth into them. In making the large metropolitan sewer near Dulwich, they have, during the last few weeks, found fine specimens of fossil Teredo.

I mentioned my idea, in which he agreed with me, that Pholas bores entirely by mechanical means, to Mr. Woodward, and told him that I had just been endeavoring to make castings, by pouring in hot wax, of the rifleing marks inside the borings of the Pholades from the Brighton chalk, and that I had partially succeeded; when he pointed out to me, in a case (just under the nose of Megatherium skeleton, at the end of the long North Gallery), a most beautiful *natural* cast of these rifleing marks. The animal having been gone from the sole, iron sand in a soft state has filled it up, and has afterwards become as hard as an ordinary hammer. The markings on the side of the sole, made by the shell, are beautifully cast, and stand up in bold relief. They remind one much of that instrument which the cook uses to make lines and patterns upon soft pie-crust. This instrument has a sort of spur rowel at its end, and the projections on this rowel are much like those on this natural cast of the boring marks of *Pholas crispatus*. The Rev. J. Layton presented this remarkable specimen to the Museum, and I, for one of "the public," thank him for so doing. It was found at Harboro', Norfolk.

Mr. Woodward also shewed me a mass of hard mica slate rock with a gigantic Pholas in it, and a lump of gum from Australia; probably floated down some river into the sea; also bored by a Pholas.

The breakwater at Plymouth is, I hear, much injured by the attacks of the Pholades. I have seen a beautifully polished specimen, showing their ravages in the stone of which the breakwater is formed. It is apparently a limestone, and is as hard as marble to the touch.

It is, I believe, a well authenticated fact, that the Pholas will not bore into Portland stone (of which the Cathedral of St. Paul's is built), on account of the quantity of silica or flinty matter it contains.

The new breakwater at Weymouth is being made of this stone. The convicts who quarry it in the island above, load railway-trucks with huge blocks of it; these run down an incline. It is a work which redounds to the great credit of Mr. Coode, the Engineer; and this gentleman may congratulate himself that there are not other "Engineers" at work, in the shape of boring shells, to eat into the foundations of his breakwater.

My father collected what information he could on the subject of Portland stone not liable to boring shells; and Mr. Roberts, who then lived at Lyme Regis, in Dorsetshire, made observations for him. Mr. Roberts writes: "There is no block of Portland stone at Lyme Regis older than 1793. *No Pholas has pierced any Portland block at all.*"

But have we no shells on land which act the same part on land in boring *stone* as does the Pholas in water? My father was, I believe, the first to point out holes in old stone walls and other edifices made by *land snails;* and, in 1841, he published in "The Proceedings of the Geological Society," p. 2, a paper upon the agency of land snails in corroding and making deep excavations in compact limestones.

This paper runs as follows:—

"During the meeting of the Geological Society of France at Boulogne, in September, 1839, Dr. Buckland's attention was called by Mr. Greenhough to a congeries of peculiar hollows on the under surface of a ledge of carboniferous limestone rocks. They resembled, at first sight, the excavations made by Pholades; but, as he found in them a large number of the shells of *Helix Aspersá*, he inferred that the cavities had been formed by snails, and that probably many generations had contributed to produce them."

A few years since, the Rev. N. Stapleton informed the author that he had discovered at Tenby, in the carboniferous limestone on which the ruins of the castle stand, perforations of Pholades, thirty or forty feet above high-water mark; but, having recently examined the spot, Dr. Buckland ascertained that these excavations were the work of the same species of Helix as that which formed the cavities in the limestone near Boulogne, and he found within them specimens of the dead shells as well as of the living.

That the perforations both at Boulogne and Tenby were not the work of Pholades, Dr. Buckland says is evident.

First, from their size and shape, which, instead of the straight and regular form, accurately fitting the shell of the animal by which each hole was perforated, are tortuous, irregularly enlarging and contracting, and rarely continuous in a straight line. The holes, moreover, are often separated by only a thin partition, or are confluent.

Secondly, because they are wanting on the upper surface of the projecting ledges of limestone; whilst, on the sides and lower surfaces of the ledges, they are excavated to considerable depths.

The above reasons, Dr. Buckland says, against the excavations having been made by any marine lithophagous animal, are favorable to the hypothesis which refers the production of them to snails.

"These animals," he observes, "could find shelter only on the margin or lower surfaces of the projecting rock, and the irregular form of the confluent cavities correspond with that of the clusters of snails in their ordinary habitat and hybernation; and if to these reasons be added the fact of finding both living and dead shells in the excavations, the evidence, the author conceives, is decisive as to the agency of snails in producing the phenomenon under consideration.

"In conclusion, the author offers some remarks on the means by which these hollows have been corroded; having been overlooked in consequence, he suggests, of their having been probably referred to the action of the weather, or to original irregularities in the composition of the stone."

My father intended to pursue these investigations further, in order to ascertain whether the cavities were hollowed out by these snails by means of an acid secreted by them, or by means of their rasp-like tongues; like sea-limpets, most probably the latter. This investigation I propose to follow up myself, if it has not already been done by some more competent person.

When wandering along the sea-shore at Brighton, I picked up a mass of black and half-decayed-looking wood; through this, too, the Pholas had made its hole, and a fine patriarch Pholas he must have been, for his house was big enough to admit two fingers. But what was the wood, and whence came it? Some boatmen had but a few moments before—complaining of the hardness of the times, the coldness of the weather, &c.—been begging for alms to buy fuel: why did they not pick up this black-looking wood, of which there was abundance thrown up on the beach? Probably, I thought, because it won't burn; so I took a bit home, and put it into the fire. As I expected, the wood would not burn; it crumbled and smouldered away slowly into powder, emitting at the same time, a disagreeable smell, like sulphur and bitumen being burnt together. A section made with the knife into it exposed a beautiful woody fibre. I therefore pronounced this wood to be lignite, or wood in a state intermediate between fresh wood and coal. As there were such quantities of it about, it must be certain that there must be a bed of lignite under water, not far from the shore. This idea was shortly afterwards confirmed by a very intelligent man, who was found mending his nets; he informed us that there was a bed of it not far off, to the westward of Brighton, and that they called the place where it was found the "Jenny ground." Why it was thus named he did not know, but he did know that there were plenty of whiting to be caught thereabouts. The wood he called "*Strombolo.*" The reason of this was also unknown to him; but, upon consideration, I thought it likely that it was so christened by some sailor who had seen the mountain *Stromboli*, and who had inhaled the sulphurous vapors common thereabouts. The smell of the wood had recalled to his mind the mountain, and he had named it accordingly. This strombolo was bored through and through by the Pholas, and a fine time of it they must have had on the "Jenny ground," boring away to their hearts' content. Upon mentioning my idea of the derivation of the word strombolo to my friend, Mr. Roberts of Coved, he kindly sent me the following communication:—

"Strombolo has been erroneously supposed to have some reference to the volcano. It means in

Flemish, stream-balls, ström-bol, and was so named by some Flemings, who settled years ago at Brighton. Between Kent and Sussex and France, are beds of lignite. After much agitation, pieces of this black material, laminated with a transverse fracture, floats up and is driven ashore, when it is water-worn and rounded,—hence stream-balls. A bye-law was passed at Brighton against the burning of strombolo, as the smell emitted is very unpleasant. Strombolo is to our channel much like amber in the Baltic. It floats up and drifts ashore, from the brown coal-beds of the bottom of that sea."

In the Museum at Oxford are some very pretty specimens of fossil wood. The wood had been bored when in a recent state by the Pholas, and had subsequently become, by the infiltration of silica, converted into a fossil. The borings are, in many places, still wide open, and filled with most beautiful crystals, looking like a fairy's grotto.

Submerged forests are not uncommon in some parts of our coasts. In 1838 Mr. Hall, of Woodside, near Liverpool, stated the following curious fact:—

"By order of the Corporation, a sewer was made at a place called Beacon's Gutter, the boundary line of Liverpool, on the north shore; and the men in digging found a great quantity of hazel nuts, imbedded in a sort of peat, formed of decayed wood. They were found about eight or ten feet below the surface. Between high and low water mark large pieces of oak have been found in a very good state of preservation, the interior being quite hard, and fit for fire-wood.—Many of the nuts appear to have been gnawed by squirrels or mice; some of them crumble to pieces on being touched; but the shell of a great many is tolerably firm. Large quantities are in possession of different individuals."

There is also a submerged forest off the coast of Hastings, and ornamental boxes, &c. are made of this semi-fossil wood.

In Sir Henry Delabeche's "Geological Observer," several of these submerged forests are mentioned. Sir Henry writes—

"Various nooks and corners of oceanic bays, where we may suppose vegetation could have flourished under differences of level, so that more dry land was exposed, should be examined, as well as very sheltered situations, in places less open to the ravages of the sea. Thus a part of the coast of Tiree, Hebrides, and of another in the Bay of Skaill, mainland of Orkney, though both exposed to the ocean, furnish the remains of these 'forests,' as well as the ramifications of old estuaries, amid the shores of the British Channel and Severn. The forest (in this place,) passes beneath a considerable portion of the flat low land, commonly known as the Bridgewater levels. Dr. Buckland and Mr. Conybeare mention oak, fir, and willow trees, sometimes of large dimensions, partly rooted as they grew, and partly prostrate, fifteen to twenty feet beneath the surface of the Bridgewater levels. Furze and hazel-bushes, with their nuts, are intermingled with them."

Occasionally the bones of quadrupeds and the traces of their foot-prints are discovered in these forests, as also the remains of insects, which are important, as enabling the observer to consider the distribution of the terrestrial animal life, as well as of the plants of the time. It lies among the vegetable accumulations apparently of this date. On the banks of the Humber, remains of the red deer and the fallow deer have been detected, and in the "submarine forest" of Minehead, Somersetshire, the bones and antlers of the red deer are discovered amid the upright stumps of trees (chiefly oaks) now below the level of the sea, and covered by it at high water—the trees rooted as they grew. The latter is especially an interesting circumstance, as the red deer are still found wild in the adjoining forest of Exmoor, so that the change of level has been effected since the red deer inhabited the district.

Extending our researches into Cornwall, we find that a change of level may have happened, submerging vegetation in its place of growth, even after the introduction of man into Western England; for at the Carrion Tin Steam Works, north of Falmouth, human skulls are stated to have been discovered among the trees, and other vegetable remains, covering the stanniferous gravel. These forests were tenanted also by at least one species of large quadruped, which is now extinct. Among other places where these are found on the shores in that district, there is a considerable tract of low ground, extending from the mouth of the Neath river, eastward beyond Port Talbot, fringed by a covering of low sand hills. On the surface of the clay in which the trees are rooted, footprints have been here and there detected, as if in passages amid the trees, by which animals found their way through them—these footsteps are of various forms and sizes, some clearly those of deer, while now and then a large impression would be observed, resembling that of some gigantic ox, even of the largest size now known.

The "submerged forest" at Brighton, judging from the quantities of bored fragments of wood we pick up, must swarm with Pholas, and I was still further convinced of the fact when I inspected the chain-pier.

The piles of the pier are formed of wood, and short work of this wood the Pholas would make, if they once established a colony on it. They would pierce it in all directions, till it would become as full of holes as an old target of a well-patronized toxophilite society, and then "down would come Pholas, chain-pier and all." Not so, however. Sir S. Brown, the architect of the pier, has stolen a march upon his little hungry enemies, and has circumvented their cunning. He knew well enough that their boring apparatus, though it would cut wood and stone, would not cut iron. To enclose each pile of the pier in a solid case of iron would not only have been expensive, but also clumsy looking, so he has hit upon a much more simple plan; he has had made many thousand iron nails, with short spikes, but with enormously broad heads. These nails have been driven into every portion of the wood which is ever liable to be covered with water, as close together as it is possible to drive them; so that no portion of the wood whatever is exposed, and the great piles look as though they were covered with a coat of mail, such as we see worn by the heavy cavalry who bestride their wooden coach horses in the Tower of London.

From the Brighton architect's plan of arming his piles with nails, we too, in these burglarious times, may learn a lesson—not that we have Pholas to contend against, but another species of boring animal, namely, the house-breaker. This creature is nocturnal in his habits; when the inmates of the house are fast asleep, he quietly applies his boring apparatus—a "centre-bit," well oiled—to the softest place of the house he can find, namely, the wooden door; a few swift turns, and there is a Pholas-like hole big enough to admit one's finger; a few more holes like it, and a circular portion is bored out of the door, big enough to admit his hand, which said hand instinctively finds out the lock, undoes the latch, and the burglar enters. Now luckily for us, this "centre-bit," like the boring apparatus of the Pholas, cannot touch iron. We, therefore, advise all householders to go to the expense of a few shillings, and to nail broad-headed iron nails, with longish spikes to them, thickly all over the *inside* of their doors and shutters; *outside*, they would not only be unornamental, but useless, for they would tell the burglar that his visit was expected. Inside, however, they are not very readily seen. When the centre-bit from the outside touches the spikes, it will snap like a bit of glass; and even should twenty centre-bits be applied, they will meet with the same fate from all parts of the door, to the great discomfort of the biped Pholas.

Besides Pholas, there are other kinds of wood-boring animals living in the sea. Teredo (from the Greek TEREO, I bore,) and Zyloplaga, or the wood-eater, are the most to be dreaded. At the Institution of Civil Engineers, December, 1849, a discussion was held on Mr. Paton's description of the Southend Pier, and the vagaries of the Teredo Navalis. The following is an abstract of this paper:—

"Numerous specimens were exhibited, and commented on, of timber thoroughly perforated by worms; whilst, beside them, under the same circumstances, the Jarrow wood, from Australia, was shown to have remained completely free from injury.

"The reference to the age of Homer, as an instance of the ancient ravaging habits of the Teredo, induced a return to Geological questions; and it was shown, that in the London clay, remains had repeatedly been found, of timber perforated by sea-worms. The Oolite and Greensand formations also exhibited petrified wood, filled with boring Mollusca.

"The early state of the Teredo was noticed; when escaping from the egg, in the shape of a free swimmer it was drifted about with the tide until it met with a log, a pile, or the side of a ship, to which it attached itself, and making an inroad into it, became a non-locomotive animal of different form and habits, never again to leave the habitation it had burrowed for itself in the body of the timber.

"The various materials, such as Kyan's Corrosive Sublimate of Mercury, Sir W. Burnett's

Chloride of Zinc, Margary's Salts of Metals, Payne's combination of Muriate of Lime and Sulphate of Iron, forming in the timber an insoluble compound, and Bethell's Creosote, or Oil of Coal Tar, were discussed. All had their partisans, and were stated to have succeeded and failed under certain circumstances. Specimens of piles from Lowestoff harbor, whose waters were notoriously full of worm, showed that timber in a natural state was, in a few months, thoroughly perforated by Teredo in the centre, and Limnoria on the surface ; but that piles, which had been properly saturated according to Bethell's system, in exhausted receivers, and subjected to such pressure as insured the absorption of about ten pounds' weight of the creosote, or oil of coal tar, by each cubic foot of the timber, were perfectly preserved from attacks of marine animals of any kind.

" In one instance a partially creosoted pile had a notch cut into it, deeper than the impregnation extended; a Teredo made its entry, and was found to have worked in every direction, until it arrived within the reach of the cresote, when the animal turned away and eventually left the pile.

"Bethell's system was admitted, by all the speakers, to be that which hitherto, after many years' experience, had afforded the most satisfactory results.

"Some most conclusive experiments, instituted by Mr. Rendel at Southampton, were stated to have produced the same results; and at Leith all the piles were weighed before and after their saturation, to insure their absorbing the full allowance of at least ten pounds per cubic foot."

I believe, however, that the plan of using broad-headed nails, as at Brighton, is better than any chemical preparation, and is now generally adopted in English harbors.

In the British Museum, Eastern Zoological Gallery, at the end of table-case, No. 30, and close by the stuffed herons and cranes, is a very beautiful series of wood, bored by Pholas and Teredo: there may be noticed a specimen of lignite, similar to what we find at Brighton, with the boring shell still remaining in the hole it has cut for itself.

The following specimens are also placed in the cabinet, and are well worth attention, as showing what serious mischief these boring animals do to the woodwork of our harbors, and even to the ships themselves. Yarmouth seems especially to suffer, for we find "a section of Quebec elm, in the sea four years, and English oak, both from the Yarmouth pier, and both quite honey-combed with the 'worm,' as the shipwrights call it." Then we have the following bits of worm-attacked ships:—"Portion of bottom of her Majesty's ship *Blenheim*, 74, lost, with her commander, Sir Thomas Troubridge, and her crew, on her passage from India. Plank of the barge of her Majesty's ship *Etna*, employed on survey of coast of Africa, 1832—one hundred days in the water, bored by Teredo Navalis."

In the Royal College of Surgeons is a very fine specimen of the keel of a ship, bored by

Teredo; it is so fragile that it is put under a glass case.

I learn from an officer, who served in the yacht *Fox*, (the little vessel whose name will ever be connected with that of Sir John Franklin), that the *Fox* had no copper on her bottom when she made her perilous voyage; and that the whale-ships are not protected under the water line. Hence I conclude, that there are no boring Pholades or Teredo in the cold waters of the Arctic Seas.

I have subsequently looked carefully over the "Franklin Relics," and among all the bits of wood, both small and great, there preserved, I can find no trace of the boring of Teredo or Pholas.

It is the Teredo which did so much mischief to the Russian ships which were sunk before the mouth of the harbor of Sebastopol. My friend Mr. Roberts has published his remarks, in 1857, upon this subject, in *Household Words* which I quote *in extenso*, as they are the result of much patient investigation.

"When the Russian commander of Sebastopol found that he would not be able to screen his ships from the fire of the enemy, and that the fleets of England and France would come into the harbor, he sunk the great fleet of sixty-two ships! The British sailor sighed as he viewed the tops of the masts peeping out of the water, and counted the loss this act was to himself. What a rich prize did the harbor of Sebastopol contain!

"After the city had fallen, a company of divers, under Mr. Deane, was sent out from Kent. The director of the party was prepared to send down his men, and furnish a report of the condition and situation of the ships; but the guns from the north side prevented the vessel, which bore the diving apparatus; and then peace came, and the sunken ships, that cost millions in their construction, were left to the Russians. They have not been raised; though a contract has been entered into with an American, who is reported to have shown great skill in recovering from a depth large sunken ships in other parts of the world. A newspaper account has conveyed to the public many particulars of the intended plans, and of the descent of a diver to view many of the costly ships. This explorer found the American raisers had been anticipated by a more numerous, indeed innumerable party of joint carpenters and masons, destroyers. Let us confine our attention for a while to these operatives, whether working for themselves alone, or revenging the Turks for the affair of Sinope.

"The boring worms are most destructive creatures. Like other pests, as man calls such things, when he suffers individually by them, they are of great use in the world. Their mission is well defined. They are of several sizes, and are generally spoken of as the Teredo or boring-worm. There is a smaller kind, which is very destructive, the Limnoria terebrans. In many climes the rainy seasons cause floods, which bring down and lodge at the mouths of the rivers thousands of trees, which threaten to close their mouths. This would be the inevitable result if the tree were left to undergo slow decay. The Teredo, however, comes to the aid of man, and renders incalculable services in boring every tree till it is internally like a honeycomb, and breaks up and floats away piecemeal. Thus an entrance is preserved and an outlet maintained for a country extending perhaps for hundreds of miles. In civilised countries the services of this great family are not required. The inconveniences and the costly damages they effect are the occasion of loss, and of a serious expenditure to prevent their ravages. It has been estimated, that at Plymouth and Devonport alone the boring-worms have in one year damaged Government works to the amount of eight thousand pounds. In 1731–2 they committed such ravages in the piles forming the sea-defences of Holland, that the Dutch were seriously alarmed. In England, when oak timber was a drug, this wood was much used for marine constructions, such as harbors, groynes, &c. In Queen Elizabeth's reign the name applied to them in petitions setting forth the losses sustained, was expressive. This was Artes.

"The animal of the Teredo is like a long white worm, varying from a foot to two feet and a half in length, and about the size of a person's finger. Mr. Brunell, perceiving how this soft creature bored on, and encased itself in a calcareous piece of masonry or tube as it progresses, perceived how he might bore the great Tunnel under the Thames. Men drove rods into the mud from a shield, which was moved forward as they bored their way, and a brick arch was constructed behind, in imitation of the calcareous tube of the worm. Thus does observant man treasure up and apply what even the animals in the lower scale of creation can teach.

"The destructive Teredo, like the lion, has his jackal—the Limnoria terebrans, or gribble worm. Woodwork in most situations, as posts in harbors, and piles of wooden bridges, must be protected by copper sheathing or square-headed nails made for the purpose. The gribble finds some little space, bores in and destroys the wood around. The Teredo then finds an entry, and destruction follows. The wooden bridge over the estuary of the Teign was destroyed some years ago. Other similar works, and particularly projecting landing piers, have been either eaten away or jeopardised.

"In the account that has been circulated, through the medium of the press, of the sunken fleet, its condition, and the ravages of the worm, there is matter that is not intelligible to one who studies the habits of the lower animals of creation.

"We are told that a diver has gone down to visit the great fleet which he finds in the middle of the harbor, and upon the north side, lying there on the sand; on the south side, on mud.

"It is further stated that the depth of water is sixty feet. Now this is a very convenient depth, for Man and the Teredo are limited in their operations to the same depth from the surface. A ship sunk in water twenty fathoms, or one hundred and twenty feet deep, is safe from the family of carpenter-mason worms, of whatever species, (and for which consult that admirable treatise, the Manual of the Mollusca, by S. P. Woodward, Esq., F.G.S., of the British Museum). Man has gone down so deep, but he can do but little under such a pressure; and that is his limit. Perhaps few men could descend so far. At half that depth each of Mr. Deane's party with his crowbar is in power as a giant, and in the four hours during all which time each remained below, working at a wreck, performed prodigies!

"The Russian ships that lie upon the sand are reported to be untouched by the worm; while those ships upon the mud are in this short time so much affected that they will be worthless.

"Reading this statement, many understand that the Teredines do not exist where the bottom is sand; and where the mud is, they are ready to transfer themselves, like rats in a dock, at once to do execution upon sunken timber fully-grown and with fully-developed powers.

"From my experience I can readily believe there are seasons when the ravages of these animals are more felt than at others. As a member of a south-western town-council, I learnt that elm sheathing to masonry or oak-posts were, though upon the sand, at times eaten up presently.

"As to the mud of Sebastopol harbor being the habitat of the fully grown worm, this must be quite a mistake.

"The balk, floating planks, and beams of different woods, occupied and quite honeycombed by the Teredo, show no outward signs of being tenanted by the carpenter-mason. The worm entered when very minute, when an emigrant upon the look-out for a domicile, and fully endued with the powers of locomotion. Some bored for three feet and encased themselves with masonry as they proceeded. No one interferes with another, in which admirable social excellence they are probably guided by sound. They work with the grain and are not afterwards migratory. Like Charles the Second, they have no disposition to go again upon their travels.

"A mere view would not have allowed the

diver to judge of the ravages already effected. It may be that he cut off pieces of the ships, and so ascertained with some precision.

"The venerable line-of-battle ship sunk in Torbay, when getting under weigh, some seventy years ago, was eaten to pieces. When there was motion, then pieces would break off and float up, and fish by shoals remain around expectant of their meal. There is no hooking ground for fishermen so good as that round a sunken ship."

Being convinced that such an important subject to the naval architect as the boring of Teredo could not have escaped the notice of the officers of that most interesting institution, the Royal Service Institution, Whitehall-yard (of which I have the honor to be a member, and which I advise everybody to visit), I went carefully over the museum and found on the wall (close to the Nelson relics) a fine specimen of Teredo-eaten wood from the Sebastopol ships. The reader may, therefore, with his own eyes, see the effects produced by the worm as described by Mr. Roberts.

The ravages of the worm are most apparent, and it will be easily imagined that it is impossible to use the ships even if they could again be raised from the bottom.

Close underneath these two interesting specimens is a most curious relic ; it is :

> " Gun from the wreck the *Mary Rose*, sunk in action with the French off Spithead. Time,—Henry viii. 1545.

The wood-work in which this is placed is, I believe, part and parcel of the gun. I was much delighted to find the large end of this also bored by the Pholas shell but not by Teredo. And here I may remark that the worm-like Teredo always makes his tunnels regular, and never cuts or crosses into that of his neighbor ; whereas Pholas is not so polite, and gouges the wood he attacks in all directions.

From this ancient gun I also learnt a useful lesson, and Mr. Brett, of submarine telegraph notoriety, confirmed my observation ; it is that the oxide, &c. formed by the rust of iron under water cements to itself all the shingle in the neighborhood; and upon this gun pieces of shingle are seen adherent from this cause. Outside the museum, in the open air, a much better example of this can be seen; it is an anchor from the wreck of one of the ships which formed the Spanish Armada; it was sunk off the coast of Ireland in 1588, and is quite covered with a mass of shingle and shells. The dining tables in the boys' hall at Westminster are said to be made out of the oak wood from the Spanish Armada; they are beautifully polished, and will last many hundred years more. In one of the tables Mr. Cleghorn the butler shows a round hole from a cannon shot; to my mind it looks more like a natural knot in the wood. Now this fact of decomposing iron making a natural cement is most important to the engineer who lays submarine telegraphs (it has been doubted by some), and not a little assists the preservation of the expensive and carefully prepared wire.

For my own part I believe they will never be able to recover the great Atlantic cable, it having been probably firmly cemented to the surrounding shingle before this time.

As being at present resident in London, I had not the opportunity of examining personally the action of "the worm" upon the bottoms of ships, I paid a visit to an establishment where I could see these planks at my ease. I accordingly introduced myself to the son of Mr. Castle, Baltic wharf, Milbank, close to Vauxhall Bridge. Mr. Castle buys old ships and breaks them up for firewood, which he sells at a cheap rate, and capital fires they make. One gentleman, he tells me, has had a set of furniture made from the oak planks of these old ships. On the wall outside the premises are fixed the figure-heads of some of the ships that have been broken up; they are more or less artistically carved, and therefore worth preserving: and here we have a curious collection, such as George IV., Neptune, Queen Adelaide, Acheron, W. Manning, Jupiter, and Lord Sidmouth. Over the door of the *counting-*

house is an immense wooden arm holding out a money bag which is gilt; it belonged to the figure-head of the ship *Commerce*, and is not a bad emblem of the office over which it is fixed.

Mr. Castle kindly showed me over the yard, having previously taken from his desk a fine specimen of Teredo-bored wood which he had preserved from one of his ships. One of the workmen informed me that they found "the worm" in old ships, and generally in ships that had been sailing in tropical climates. It prefers African oak and teak, for "them things has as nice appetites as we have ourselves." Last year he found in the *Flamer* and the *Bathurst*, which were broken up, a great number of "them short worms with hard heads that makes long holes and lines them with cement as they go along" (a capital description of Teredo). "These two ships were quite full of them; they scrunched under the feet, and the birds came and picked them out to eat; they will get in anywhere where the copper is knocked off the bottom of the ship, and we finds 'em principally on the 'flats' of the vessel; when they comes to an iron or copper bolt they turns a one side and goes round it, as they don't like it." This man promised to save for me the next good specimen he obtained. I observed in several portions of wood among the bones and skeletons of the ships that were about the yard, a most interesting natural preservative process which takes place in oak ships which are fastened together with iron. I found many bits of oak-wood stained of a blue ink-like color. Here then is the explanation; the iron of the bolt becomes decomposed by the action of the water, and combines with the tannic acid in the oak, thereby forming, as everybody knows, *genuine ink.* The wood saturated with this ink resists the action of the water better than the uninked wood, and the worm will not bore into it. I obtained several fine specimens of this wood; the stained part tastes bitter, exactly like ink, when crushed between the teeth. This same phenomenon may be observed in oak gate-posts in the country which have iron fastenings or nails driven into them.

But although, at Brighton, the burglarious Pholas is kept out of the wood of the pier by means of the iron nails, yet other parasites have established themselves on the iron armor thereof; for I observed large masses of the common mussel-shell hanging on to them, like huge bunches of black Hamburgh grapes. While I was looking from them above, a colossal wave came thundering in, giving the poor mussels a tremendous blow, and making them rattle again.

Why did not that wave, I thought, knock off the shells? An opportunity soon came, for, on examination, I found that they were all strung on together, like a rope of onions. There is a centre rope, upon which, or rather to which, each shell is attached by an independent string; nay, more than this, each shell has tied himself on to his neighbor for common safety's sake, like the travellers do going up Mont Blanc, as we see in Albert Smith's picture. The shells' rope looks like a very coarse bit of unravelled tow which has been soaked in water; it is, in fact, a sort of natural string. I have seen a ladies' glove made from the tow, or rather silk, secreted by the pinna-shell, and the pinna uses his silk for the same purpose as the mussel his tow, namely, to anchor himself firmly to a rock. In the British Museum may be seen some exceedingly fine specimens of the pinna-shell, and also gloves made from its silk or byssus.

On the ends of the iron drain-pipes at Brighton, I found another cluster of these mussels. Here, the object to which they had attached themselves being round, they had moulded themselves accordingly; and, instead of being in the form of a bunch of grapes, they had closely and firmly spun themselves together, so as to form a thick, firm sheet, which could be easily peeled off, being nearly as firm as a house-door mat.

This fact of their binding powers has been observed and acted on by the French engineers in Cherbourg; for, to make the breakwater, they have planted, as it were, several tons of mussels, throwing them upon the loose masses of stone. In course of time, these little workmen will spin their string-like webs for self-security's sake, and will bind the loose stones firmly together, thus unconsciously making a living cement, more durable than any material ever invented by man. Thus we see that these two apparently unimportant shells—the Pholas the destroyer, and the Mussel the preserver—silently and unobserved in the depths of the ocean, may bring about important changes in the affairs of men, and even turn the scale in the destiny of nations.

I must now, with the reader's permission, change the scene of the "Hunt on the Sea-shore" to that charming sea-side town, Folkestone, where I had spent most of the autumn of 1859, and much of the following has appeared in *The Field*, under the "Naturalist" columns, and by the title of "Sea-side Gatherings."

Folkestone, anciently called "Fulchestan" and Folkston," was of considerable note even so long ago as the times of the Romans. As we have a "King's-stone" near London, so we have a "Lapis Populi," or people's (folks) stone in Kent; and from this we may conclude that, when the invading Romans came over to this island, they found numerous inhabitants at this town. It is believed to have been the site of one of their numerous fortresses, which they wisely erected to guard themselves against the bugbears of those days—the Saxons. Just as *we*, the present occupants of this island, have a military station at Shorncliffe, for defensive purposes, on the heights above Folkestone, so did the Romans form a camp and look out upon the top of Castle Hill, about a mile and a half behind the town, and just above our present encampment. There is about as much trace left of the Roman camp now as there will be of Shorncliffe a thousand years hence, viz. a few earthworks covered over with nature's carpet, green grass and turf. But if the works of man are perishable, the works of nature are not.

The action of the waves upon hard stone near this town, as well as upon chalk at Brighton, can be well observed; for, in order to preserve the town from the white-crested battering-rams of Father Neptune, two sea-walls, called "The Horn," have been constructed, on each side of the entrance to the harbor, with stones collected from the neighboring cliff. This cliff is composed of sandstone, disposed in layers, which crack off in places from each other like the sheets of an unbound book. There can be no doubt that this sandstone, now as hard as sheet iron, once formed an ancient sea-beach, and that, long before the creation of man, the waves rolled in, and the tides ebbed and flowed upon this now fossil sand, exactly as they do at the present moment. Examine it, and you will find well-marked holes and borings of the "lugs," or sea-worms, the decen dants of which are this day alive in hundreds in the mud of the harbor not three yards off. When the lug makes his boring, he lines it with a sort of cement, and thus leaves a regular open tunnel; the next tide fills up these holes with sand and mud, and we find these turned into fossil, resembling holes made with a common walking-stick. These ancient lugs were evidently gigantic fellows, if we judge by the holes they made. They themselves have disappeared, but their holes remain, converted by the magic hand of Time into adamantine rock.

We observe in this formation a good example of the tendency of sandstone to assume rude crystalline forms, for it often presents blocks the edges of which are rounded and smoothed with almost mathematical accuracy ; and this explains how a large stone as round as a cannon-ball, and exceedingly like one in appearance, was obtained by the coastguard man at Sandgate (a village close by) to form the summit of his garden gate-post. At first, I thought it was some ancient stone cannon-ball of gigantic dimensions ; but, no, it was never manufactured by the hand of man, but by the ocean, whose delight it seems to be to transform everything submitted to the action of her workmen—the waves—into a rounded form. Who ever picked up a stone on the sea-shore, that had been long submitted to the action of the waves, that presented any edges or square corners ? The ocean has put these blocks of semi-crystallize sandstone into her water lathes and converted them into cannon-balls. It would be entering into a difficult geological question as to *when* this sandstone formed the actual sea-beach. Suffice it to say, that it was long anterior to even the creation of man, that it had nothing to do with the deluge, and that there is no doubt that a gradual upheaval of the land has taken place at some very remote period, raising up the chalk downs, and the various strata near it.

In the hard stone blocks which compose the walls of the "Horn" are thousands of the holes of the Pholas, made at the time that these stones were lying in the sea, at the base of the neighboring cliff; and when the Pholades die, and their shells drop out, numerous sea creatures use their holes for habitation. Little crabs, shrimps, sand-hoppers or (as the men call them here) skip-jacks, together with thousands of little animals exactly like common woodlice, which the amphibious boys about the pier call "monkey-peas" (and not a bad name either—monkey, because of their activity; pease, because they roll up, and look like pease), are seen swarming in every direction. These little "monkey-peas" are vegetable fee ders, and they will get into the boats and devour the planks forming their sides, if not well looked after. They also eat up the sails and nets, if left rolled up any time. Then, again, the whole surface of the rocks are coated with hard white peppercorns. These are thousands of acorn shells, or barnacle shells—stupid and uninteresting things when left high and dry, but beautiful when seen feeding in a vivarium.

But have all these little beasts no enemies? The long swell of the tide brings them in in legions: first of all come sailing along the jellyfish, curious things, that look like a flexible soup-plate of calves'-foot jelly. They are of a bell shape, and swim along by means of contracting and expanding their bodies, like the opening and shutting of an umbrella. They have numerous queer names. The Folkstone name for them is "slutters;" at Dover they call them "water-galls" and "miller's eyes," and "sea starch;" at Portsmouth they are called "blubbers;" elsewhere "slobs" and "slobbers." When preserved in the cabinet, they are placed among the Medusæ, or

Acephalous Mollusks. They are great bores to the fishermen, who often catch so many in their trawl nets that they can hardly haul them on board their boats. When the nets are brought up, they appear to have been soaked in a strong solution of common starch, and require boiling before they are fit for use again. There is also a species of jelly fish that has long strings like wire tendrils hanging down from them, which have the power of stinging: these are called "sea nettles" and "stingers." These latter kind are not common in shore; their home seems to be out at sea, where they are caught of an enormous size. Hard weather drives them in, and then let the bathers beware. I know a young lady who was severely injured by one of them: being ignorant of its powers, she foolishly caught it as it floated by, put it to her mouth, and bit it; she suffered acute pain and swelling of the lips and face for several days, in consequence of the sting she received from the brute. The sensation of the sting is like a sharp blow with a bunch of common stinging-nettles; it causes a sense of irritation and itching all over the body. On the 25th of August, when out fishing, I observed innumerable specimens of a very large kind of jelly-fish, which the fishermen call "blue slutters." They were about the size of a very large cabbage, and presented round the margins of their cup-shaped bodies an edging of a most beautiful blue color. I dissected several on the spot, but was soon obliged to desist; for the left hand (in which I held them) became covered with a red rash, and the fingers felt numb and as though seized with an attack of "pins and needles." It is, doubtless, this power of stinging that enables the slutters to procure food. The fishermen declared that they subsist on the water, but I found out they ate something more substantial. I never examined a specimen without finding in one of four cavities which form the stomachs several skip-jacks, some alive, some half digested. Sometimes I have found these little brutes so recently made prisoners, that Mr. Jack has made a skip out of the jailor's stomach, and was free again in the sea.

I now understand why they are called "miller's eyes." When the stomach is full of food, they present the appearance of having four great goggle-eyes staring through a pair of spectacles. I know not what creatures eat these "slutters" when alive, but I found one dead, around which were a whole shoal of hungry shrimps, eating away as hard as they could. When the "slutter" is dried, it shrivels away to nothing, there being so much water in its composition. They are on this account useless to the farmer as manure. In the Chinese seas they are found of gigantic dimensions; and a gentleman informs me that when off Macao, in China, he saw the fishermen collecting them by boat-loads. They did not touch them with their hands, but caught them with large iron rakes. These Chinese (probably aware that they would dry up if exposed to the sun,) wisely mixed them up in other manure, which was afterwards dug into the rice fields.

The base of the Folkestone pier is covered with numerous plants that are commonly called sea-weeds. I do not think they ought to be called weeds, because, according to Johnson, a weed is "a noxious or useless herb;" let us rather call them sea plants," for they are quite as useful in the economy of the world of water as the green vegetation is in the well-being of the world of land. When the tide is down, these plants hang about, dirty and black looking objects. When the water comes in, they assume another aspect, and float about with graceful wave-like motions. Why are they not knocked to pieces by the restless waves? Take one example, the common "bladder-wrack;" we find every here and there, regular bladders in its substance, each containing air—and these act the part of buoys, lifting up the whole plant, and preventing injury being done by the crushing and bruising action of the rolling swell. If we make a section of these bladders we shall find inside a beautiful network of silk cobweb-like fibres, which form a pretty pattern with their complicated interlacings.—When these plants have been submitted any time to the action of the sun, the air inside the bladders becomes expanded, and they make a famous popping noise when trodden upon. Nature seems frequently to apply modifications of one and the same plan to different purposes, and man follows her example; thus we have a bubble of air inclosed in an iron case, forming a floating buoy to mark out dangerous sands, &c. to sailors, and we have a lighter kind of air, commonly called gas, inclosed in a silk case, which will float in a different medium from water, in the form of the Royal Vauxhall balloon. If we were given a yard of rope, the end of which was untwisted for a couple of inches, and were told to fasten the unstranded portion on to a rock in such a manner that the waves would not move it from its place, we should attempt our task in vain; but yet in the roots of humble and despised sea plants, which have taken their growth on bare and smooth stones, we find a wonderful example of a fastening more adhesive than any human sewing or screwing—more difficult to be moved than if it were attached by marine glue of a hundred horse power.

Among the thickets of the submarine underwood live the enemies of the crabs, the shrimps and the barnacle shells. I caught many whiting pouts, or "pouters," as they are called at Folkestone, and found, without exception, that there was a red brickdust-like substance upon the lower surface of their bodies; this felt gritty to the fingers, and I suspected it was the powdered shells of crabs. My idea was soon substantiated by my finding in the stomach of a "pouter" two or three crabs more or less digested, and their hard shells powdered by the action of the digestive process. At low tide the crabs get into the holes of the rocks, and come out again as the water rises, just in time to meet the hungry pouters, who come in with the tide to look for their dinner in the harbor.

Upon my hook, close to the pier, I caught a little black fish with a malignant, diabolical countenance. I was going to unhook him when a boy cried out, "Don't touch him, sir; he is a 'bull-rout,' and he will bite you." I shook him off the hook, and found the boy's advice to be founded on experience, for Mr. Bull-rout fastened on to a bit of wood, and bit it severely.

The bull-rout is the "goby" of Yarrell. His mouth will well repay examination—a more formidable and desperado-like set of teeth I never saw. They are more like a tiger's than a fish's teeth—but how beautifully are they adapted to their work! All the bull-routes I caught had their stomachs full of the barnacle shell, and their office seems to be expressly to feed on the barnacle. Every one knows how tightly this little shell fastens itself to the rocks, and the bull-rout's teeth seem to have been made expressly to tear it off.

Now these barnacles have a certain amount of what might be vulgarly called sense, but is in reality an instinctive act, for at certain periods I had observed at Brighton that there was a peculiar smell in the houses, as though proceeding from the drains. During a walk on the beach I discovered the reason of this. There are several iron pipes projecting out of the shingle towards the sea—at high water their mouths are covered, at low water they are open. Thus therefore the alternating smells could be accounted for—when the end of the pipe was exposed by the flowing of the water, up would come all the noxious gases formed by the drainage—when, on the contrary, the water covered the mouths of the pipes, the smells remained quiescent.

Wishing to ascertain this for a certainty, at low tide I placed my head inside one of the pipes and puffed away at a cigar. Away flew the smoke upwards into the dark regions of the pipe, and probably had it lasted long enough, would have come out in some of the smart houses facing the sea.

During the experiment, I ascertained a curious fact: the upper half of the pipe was lined with Balani or acorn shells about the size of a split pea, the same kind that we see in the Vivaria of the Zoological Gardens, and which continually expand and contract their little hand-like infusoria catchers; on the lower half of the pipe, however, not a single barnacle could be seen. The reason was evident; down the lower half was rushing a stream of fresh water. Now, fresh water is death to the Balani, and therefore they wisely keep above its level, and confine the limits of their colony to the upper half of the pipe. Snug quarters have they found inside the pipe: they get their dinner when the sea water comes up; and when the tide goes down, they have a nice airing from the fresh sea-breeze as it rushes upwards.

Upon examining the outside of the pipe, I found it to be most beautifully polished, feeling like glass to the touch. This is caused by the perpetual motion of the pebbles rolling over it; even the joints where the pipes met had their edges worn off and smoothed. In one junction, I observed that a stone had become fixed, and that the iron round it had become worn away, so that thin edges of iron were projecting over it; in fact, it was, to use a jeweller's expression, set in iron, just as precious stones are set in gold and silver. The pipe looked red; it was covered by a thin coat of rust, which is probably formed fresh every time the tide goes out. In proof of this, I adduce the fact that Robinson Crusoe-like, I observed the mark of a foot which had been set upon the pipe. This impression certainly was not that of a cannibal, but had been made by the tiny and delicate little foot of a pretty young English girl (for I afterwards saw her walking along, about a quarter of a mile away,) who had stepped upon it when passing over. The young lady had on a *new* pair of India-rubber shoes, as we dare predicate, from the fact that the impressions of the triangles usually found at the bottom of such shoes were well marked. If the shoes had been old, the triangles would have been worn off, and of course no impressions of them made on the rust of the pipe. Underneath the foot-mark, we saw the common surface of the iron with a fresh coating of rust all round; proving that, at every fall of the tide, a fresh process of oxydation or rust goes on, which, in time, must greatly diminish the thickness of the pipe.

The action of the breaking waves and rolling shingle often produces curious results upon wood submitted to their action. Long jetties of deal wood are run out from the shore at Folkestone, to enable the shingle to accumulate for the protection of the harbor; fastened upon these deal piles I observed a substance that looked like an extraneous growth. It was nothing but the softer parts of the wood hammered by the waves and shingle into a sort of pulp, which had afterwards got dry. It was arranged somewhat like the curls of the hair of an African negro, and in a pretty chequered pattern. I was much pleased to find exactly the same appearance on the board, also of deal, belonging to one of our soldiers' wives whose sick child I was attending. The perpetual friction of wet clothes, soap, &c., will therefore produce the same result upon a soft deal board, (only in a minor degree,) as do the waves of the sea upon the deal piles of the breakwater.

Cast up by the waves, and lying among the seaweed at the foot of the pier, I espied a black mass of something that looked like a bunch of burnt vial corks tied together. I asked a prawn-fisher what they were. "Sea grapes," said he. I pretended not to understand. "Well, sir, if they aint sea grapes they are tortoises' spawn." "And what are tortoises?" "Tortoises is ink spewers, and ink spewers is scuttle fish." Here, then, is the fisherman's account of the eggs of the common cuttle (not scuttle) fish. These eggs are very like a bunch of black Hamburgh grapes, each conical-shaped egg being composed of a thick, hard material, and bound by a foot-stalk on to a centre string, which is as thick as a lead pencil, and feels like india-rubber when cut with a knife. As the specimen was recently thrown up, I thought that possibly the young cuttle might be alive. I opened one of the largest eggs, and there I was pleased to find a little cuttle, the size of a split horse-bean, all alive, oh! It was perfectly in shape as its parent, but soft in substance as a bit of fresh paste made of flour. I put him in a quiet pool of sea-water, and in an instant the water became black, as though a thimbleful of writing-ink had been upset into it. The cuttle is a favorite food of many sea animals, and he has no defence but concealment. When he sees his enemy coming he has the power of emitting from his body a quantity of ink-like fluid, which obscures the water and hides his shining body. This ink, when collected and prepared, is known as sepia, and is, as everybody knows, much used for drawing and other purposes. It was most interesting to see that this little cuttle-fish, which was only half-way through his egg-state of existence, used his natural defence the very instant he first saw sunlight, and some days, may be weeks, before his proper and natural time of appearance.

Dr. Lankester, in lecturing upon the Sepia at the Royal Institution, remarked, "That like human scribes, the cuttle-fish frequently delighted to envelope itself in clouds of ink."

The ancient cuttle-fishes of geological times are found fossil, with their ink-bags quite perfect and full of ink. In my lamented father's museum at Oxford is a splendid specimen of a fossil cuttle with the ink still in it. It is mentioned in his "Bridgewater Treatise," p. 218, vol. i., and a plate given in vol. ii. He says of it: "So completely are the character and qualities of the ink retained in its fossil state, that when, in 1828, I submitted a portion of it to my friend, Sir Francis Chantrey, requesting him to try its powers as a pigment, he had prepared a drawing with a saturated portion of the fossil substance; the drawing was shown to a celebrated painter, without any information as to its origin, and he immediately pronounced it to be tinted with sepia of excellent quality, and begged to be informed by what colorman it was prepared."

Dr. Buckland, with his usual acumen, draws a conclusion from the fact that these fossil ink-bags were found full. He writes, p. 289: "I have drawings of the remains of extinct species prepared also with their own ink; with this fossil ink I might record the fact and explain the causes of its wonderful preservation. I might register the proofs of *instantaneous death* detected in these ink-bags, for they contain the fluid which the living sepia emits in the moment of alarm; and might detail further evidence of their immediate burial in the retention of the forms of these distended membranes, since they would speedily have decayed and have spilt their ink had they been exposed but a few hours to decomposition in water. The animals must, therefore, have died *suddenly*, and been *quickly* buried in the sediment that formed the strata in which their petrified ink-bags are thus preserved."

The Folkstone fishermen have observed the habit of the cuttle-fish of ejecting a coloring material when alarmed, and have christened them by the significant but not elegant name of "ink-spewers." They call them "tortoises," because, when taken out of the water, "they hunches up their backs like a tortoise."

The kind of cuttle-fish whose eggs I found has a curious shaped white bone in the centre of its soft body, and we frequently see large bottles in chemists' shops full of these cuttle-fish bones.—They are used when finely powdered for tooth powder, and also to rub out ink marks from paper—their gritty consistency makes them useful for these purposes, and whereas they are composed principally of lime, they are often cut into bits, and given to canary birds to peck at. Their beautiful laminated internal structure will well repay examination.

Besides this common cuttle-fish, there is another kind of creature very like it found in the sea near Folkestone. They have no bone in their bodies; the fishermen call them squibs and man-suckers, because they have powerful sucking discs on their long arms, with which they can take a very firm hold on intruding hands. They are not commonly found near the shore, but live more out at sea. I bargained for one, and forthwith dissected him. His length was about two feet when spread out—the body of the shape and the size of a large Jersey pear—the head in its outline much resembled that of an elephant when viewed from the front, surrounded by eight long arms, the lower surface of which were covered with numerous sucking discs, and upon turning these away the mouth was exposed, formed by two hard, black, horny beaks, exactly like the beak of a parrot—and this is well calculated to inflict a severe bite upon the unfortunate fish or crab that is held fast to it by the powerful suckers. These beaks are found fossil, and in early times of geology were a great puzzle to beginners, and there were many disputes as to what they were. Although the body seems soft and jelly-like, yet the man-sucker has a highly complicated stomach, composed of firm, muscular substance, like the gizzard of the fowl. The liver is as large as a hen-egg, and altogether it is a highly formed animal as regards its inside. The eye is very beautiful—it is oblong, like the eye of a Chinese, and the pupil is of a bright and resplendent gold color. The lens is highly complicated, and shaped like a Coddington magnifying lens.

The power of vision must be acute, for shortly after my dissection, when fishing with a line about two miles out at sea, I espied a man-sucker of a large size, floating quietly by the boat. I was preparing to plunge in to catch him, or he to catch me with his suckers, but he saw me and sank down out of sight like a ten-pound shot.—He was floating with his beak upwards, and his arms hanging downwards like the plume on a soldier's helmet, the boss of the helmet representing the mouth. Determined not to be disappointed, I offered a reward of beer, and have obtained a living specimen of a man-sucker.

The color of the specimen when first caught was that of brown silk, a pleasing color to the eye, shot with gold; but the tints gradually got dull as the animal became exhausted. Underneath the skin of the body and the arms is seen waving about as if from cell to cell a dark brown colored fluid, causing the tints of the skin to become at times almost white, at other times almost iradescent. I do not know that this fact has ever been noticed before. I find that strong brine has caused my pet man-sucker (who soon died) to turn of a red color, like the outside of a bit of boiled beef. The conical eyes are defended with eyelids, which the animal closes when the eye is touched. The natural position when at rest is with the mouth turned downwards, and the suckers applied to the bottom of the vessel—they hold so tight that the creature can be removed with difficulty; the moment one arm relaxes, the others hold firmer still. I allowed him to grasp my hand and arm—the feeling is that of a hundred tiny air-pumps applied all at once, and little round red marks are left on the skin where the suckers were applied—and when they were all fast, the animal could hardly be got off again. The sensation of being held by a man-sucker is anything but agreeable. The feeling of being held fast by a (literally) cold blooded, soulless, pitiless and voracious sea monster, almost makes one's blood run cold. I can now easily understand why the natives of the Chinese and Indian seas have such a horror of them, for in these climates they are seen large and formidable enough to be dangerous to any human beings who may be so unfortunate as to be clutched by them.

It is not impossible that the extraordinary sea-monster, the "Kraken," may have been some kind of gigantic cuttle-fish. Mr. Pennant thus writes of the eight-armed cuttle-fish:—"In the

Indian seas this species has been found of such a size as to measure twelve feet in breadth across the central part, while each arm was fifty-four feet in length ; thus making it extend from point to point one hundred and twenty feet." He further states, that "the natives of the Indian Isles, when sailing in their canoes, always take care to be provided with hatchets, in order immediately to cut off the arms of such of these animals as happen to fling them over the side of the canoe, lest they should pull it under water and sink it."

The opinion of Dr. Shaw is equally decided regarding the occurrence of this animal. "The existence of some enormously large species of the cuttle-fish tribe in the northern Indian seas can hardly be doubted; and though some accounts may have been exaggerated, yet there is sufficient cause for believing that such species may very far surpass all that are generally observed about the coasts of European seas. A northern navigator, of the name of Deris, is said, some years ago, to have lost three men in the

African seas by a monster of the colossal cuttle-fish kind, which unexpectedly made its appearance while the men were employed, during a calm, in raking the sides of the vessel. The colossal fish seized three men in its arms, and drew them under water, in spite of every effort to preserve them: the thickness of one of the arms, which was cut off in the contest, was that of the mizen-mast, and the suckers of the size of pot-lids."

In his work on the "Natural History of the Mollusca," Denys Montfort details the circumstances above alluded to by Dr. Shaw, from the account as supplied by Deris himself; and, among other instances, he mentions that at St. Malo, in the chapel of St. Thomas, there is an *ex voto*, or picture, deposited there by the crew of the vessel, in remembrance of their wonderful preservation during a similar attack off the coast of Angola (west coast of Africa). An enormous cuttle-fish suddenly threw its arms across the vessel, and was on the point of dragging it to the bottom, when the continual efforts of the crew succeeded in cutting off the tenacula with swords and hatchets. During the period of their greatest danger, they invoked the aid of St. Thomas; and being successful in freeing themselves from the dreadful opponent, on their return home they went in procession to the chapel, and offered up their thanksgiving. They also procured a painter to represent, as accurately as possible, their encounter, and the danger which at the moment threatened the termination of their existence. The above stories are taken from vol. viii. of the "Naturalist's Library," by Dr. R. Hamilton. A very striking representation is given of the Angola cuttle-fish seizing the boat in its long arms.

The Japanese have evidently a belief that the cuttle-fish will attack human beings; for in Mr. L. Oliphant's "China and Japan," I find a description of a Japanese show, which consisted of "a series of groups of figures carved in wood, the size of life, and as cleverly colored as Madame Tussaud's wax-works." One of these groups Mr. Oliphant describes as follows: "No.

5 was a group of women bathing in the sea; one of them had been caught in the folds of a cuttle-fish; the others, in alarm, were escaping, leaving their companion to her fate. The cuttle-fish was represented on a huge scale, its eyes, eyelids, and mouth, being made to move simultaneously by a man inside the head."

Mr. Beale in his "South Sea Voyage" describes an adventure he had on the Bonin Islands, where he tried to stop a large "Man Sucker," (which was four feet across its expanded arms,) from getting back into the water, the brute fastened upon his person with a murderous clutch: it had to be cut off him bit by bit.

Most of the fishermen's houses in Folkestone harbor are adorned with festoons of fish hung out to dry; some of these look like gigantic whiting. There was no head, tail, or fins to them, and I could not make out their nature without close examination. The rough skin on their reverse side told me at once that they were a species of dog-fish. I asked what they were? "Folkestone beef," was the reply.

What sort of fish is this? "That's a Rig;" and this? "that's a Huss;" and this other? "that! a 'Bull Huss;'" this bit of fin? "that's a 'Fiddler;'" and this bone? "that's the 'jaw of Uncle Owl,'" &c., &c.

Here, then, was a new nomenclature; but I determined to clear up the matter, so day after day, when waiting in the harbor for the trawl-boats to arrive, I took down my two volumes of "Yarrell's British Fishes." A class was soon assembled, and turning over the pages one by one, I asked the name of the fish whose portraits formed headings to the chapters. In this way I got a curious collection of local names. I give now only the dog-fish kind. A "rig" is the "common tope," Yarrell; a "bastard rig" is the "smooth hound," Yarrell; the "huss, or robin huss," is the small spotted dog-fish; the "bull huss," the large spotted dog-fish; the "fiddler," is the angel, or shark ray; "uncle owl's jaw," belonged to a species of skate.

I must here bear testimony to the excessive civility, and really gentleman-like conduct of the

Folkestone fishermen; at first they were shy of me, and tried to cram me with impossible stories, &c.; but we soon became the best of friends, and I really believe I have made some true friends among these rude but most honest and sterling men.

Nearly all these kinds of dog-fish above mentioned, are caught by the men who "go out after rigs" to the "hungry ground," over the Warne sands; and they catch them with "long lines," laid down all night.

Some of the large rigs are nothing more nor less than sharks of the English waters, and most formidable creatures they are. They have teeth of a triangular shape, exceeding sharp, and so arranged that if one is broken off another comes up into its place. "You see, sir, they has jaws as tears ye like a bramble-bush." The skin is not covered with scales, but with an exceedingly tough armor, which sets the teeth on edge when felt, and is "a terrible thing to dull your knife." When the rigs, &c. are caught out at sea they are thrown down to the bottom of the boat, and as they jump about there, they can be heard "grating one against each other." A rig lives longer than any fish in warm weather, but dies soonest in cold. When the lines are hauled, and there are a lot of freshly-caught rigs at the bottom of the boat, the men are obliged to be careful not to get bitten. "They all goes mad, sir, and it's like being among a lot of wild beasts." They have been known to catch hold of the men's "barbell," or fishing petticoat, and shake it. I have seen a small boat nearly full of these various kinds of fish—rigs, husses, bull-husses, fiddlers, &c. They are vagabond curs of the ocean, that go prowling and snapping anywhere and anyhow for food. The fishermen hate them because they do so much damage to the herring-nets, eating the fish actually out of the net, and often rolling themselves right up in it. At Dover, during the herring time, there are plenty of "rough dogs" and "smooth dogs," and the sea sometimes boils with them. There is a very peculiar smell about these dog-fish, and they are not good to eat boiled or fried. Ten minutes after the arrival of the boats the small fish-dealers may be seen cutting off their heads, tails, and fins, and splitting them into halves; they are then salted and hung out to dry, and taste, when broiled, "like veal chops." They are eaten by the poorer class, as "a relish for breakfast." The great heads, and the intestines, &c., are left in the harbor till picked up by the owners of crab-pots, or stalkers for baits. The livers are colled and boiled for oil for the boats in winter. There is an immense deal of gelatine in these heads and fins, and I tried in vain to persuade the men to boil them up, quoting the example of the Chinese, who esteem shark's-fin soup as a great delicacy. An Englishman is naturally a bad cook, and soup-making of any kind is not his forte.

The Rigs have a long projecting nose (one species of the shark is called "old shovel-nose," by the sailors), and this nose is of the same service to the rig as the snout is to a pig, for with it he routs about among the sand for small fish. The plaice, sole, &c., are essentially hiders in the sand, and the dog-fish comes and routs them up with his nose, and then snaps them up as they attempt to escape. I found several of these small fish in their stomachs, and in one monster the skin and the dorsal spikes of his cousin, the "pike dog-fish," called by the men, the "spur-fish." In Dorsetshire they call them "spur-dogs." In order to see how this nose acted, I pressed a rig's head into a heap of wet sand and shingle. It was quite marvellous to remark how beautifully the nose was fitted to its work; man, curiously enough, copies its shape exactly in the instrument used for paring turf. On making a section of the nose I was much struck at its internal anatomy; it is not solid, for that would be too heavy; the outer skin forms a framework, and from side to side firm white strings are stretched across so as to make a series of chambers, which are filled with a beautiful white gelatinous fluid. The organ of smell is highly developed, and the delicate membrane on which the nerves are spread (answering to the Schneiderian membrane in our own bodies) is defended by a fold of skin to keep the sand out of it, answering to our nostril. The eye, when fresh, can be compared to nothing but a cup hollowed out of a large pearl, containing a diamond as its lens, and powerful muscles move it in all directions. On sinking the head in clear sea-water, the eye glares like a cat's eye in the dark. I can fancy nothing more horrible than the glare and terrible rolling of this merciless, ravenous sea-tiger, deep down under water, and can well understand the piercing cry of horror which a human being naturally utters when he sees the murderous glance of the pur suing and man-eating shark. I heard a story that, some five years ago, the rigs for some reason migrated into shallow water somewhere at the back of the Isle of Wight, and that people "was afraid to bathe for them." Rigs' teeth are found loose on the sand of the harbor. The South-Sea islanders fasten them, in a firm and most ingenious manner, in rows on sticks, spears, &c., and make formidable weapons of them. There are several well selected and beautiful examples of this art in the United Service Institution, Whitehall. The whole of the skeleton is made of cartilage, or gristle, and easily decays; the glass-like teeth will last almost for ever; and here we have an explanation of a geological phenomenon. In many fossil-beds of England, sharks' teeth are found fossil by the thousand. In Malta they are called St. Paul's teeth. There were no fishermen to set "long lines" for them. They must have multiplied exceedingly, and when they died, and their bodies decayed, every particle would disappear but the teeth, which would sink down into the mud, there to remain till turned up by the geological hammer of the present period. All the bony material in the dog-fish tribes, seems concentrated on the outside of the body, in the form of this rough armor. Man turns this to account, for the skins of the huss, "which rasp like a file," are sold to wheel-rights, and they do certain kinds of polishing better than any sand-paper made by human hands. The fishermen collect the *heads* of these fish, and they make most convenient and handy things to scrub the wooden boards of the fish-truck. It is a curious sight to see the fishmonger's boy working away at his board with the head of a huss as a scrubbing-brush. The skins of the soles are collected and "sold for three shillings and sixpence a score to make isinglass."

On the rig lines they often catch the fiddlers; "we calls them fiddlers because they are like a fiddle." These, also, as well as the two kinds of huss, are cut open and dried for food. The fiddlers' heads remain alive a long time. I was cutting up one of them more than half an hour after its decapitation, and the jaws on a sudden spontaneously closed firmly upon my knife and held it fast.

A wandering showman came into the town, and, for the sum of one penny, I saw "the largest alligator ever imported into this country taken alive;" the thing had been stuffed the last twenty years. It had been captured on the banks of the Nile, to the great joy of the inhabitants, whose cattle and children it had frequently devoured, and its natural teeth had been replac ed with pegs of wood and tips of cows' horns;

and it was indeed "a monster." I saw also in the show "the angel-fish;" the body, and shoulders, and neck, are like a girl of fourteen; the feet and legs are like a large goose; the rest of the body like a fish: it was "caught in Africa." This wonderful thing was only a large "fiddler," tortered into something like a human shape.

Here is another wonderful story of an alligator, which is too good to be lost. It is from "the other side," as the Yankees have it.

"A dead alligator, as newspaper readers may remember, which was lately found floating in the bay of New York, awakened much speculation among naturalists; the recent discovery of the skeleton of another, almost as far north, in New Jersey, is therefore not a novelty of wonder in these regions, except from the immense size of the latter, and from a matter of most curious antiquarian interest connected with its exhumation. This last, alligator or crocodile, measuring thirty feet in length, was found at Eatontown, New Jersey, about ten miles from Long Branch, while digging for marl. The skeleton lay about six feet beneath the surface, in a stratum of green sand; underneath this skeleton, as if it had dropped from the mouldering stomach of the monster, was found an ancient coin! This coin was described to us by an intelligent correspondent, who has handled it, as about the size of a dollar; its composition, in which there is a large share of silver, being probably Corinthian brass. On the face of it is the figure of a lion, with the date '6—48' in Arabic numerals; on the reverse, amid several illegible letters, the fragmentary words, 'Arg. Procon. Latia Mo.' may be deciphered, surrounding two large letters in the centre, one of which is the Greek 'H' with an 'R' interwoven with it. We understand that the whole of this curious matter is to be brought before some antiquarian society in the regular mode, with all the necessary testimony as to facts particular of the discovery. But while those learned gentlemen are puzzling themselves as to the pro-consulship in which this coin was cast, and calling in the aid of the geologists to account for its locality, our quick-minded readers will instantly jump to the honest conclusion that this crocodile, who found his burial amid the sands of New Jersey, had, 2,000 years ago, half digested some Roman soldier in the rivers of Africa, ere he floated westward for a new meal, with the poor fellow's last coin still preserved in his maw."—*New York Literary World*, October, 1848.

While thinking about alligators and crocodiles, a letter was put into my hands from my cousin, Charles Buckland, Esq., who holds a high and responsible situation in the Civil Service of India. When he went out, I requested him to record for me all remarkable facts in natural history (and he is no bad observer) which came under his notice. He writes to me as follows:—

" It chances that I am now at Burdwan, where there is a native rajah who keeps a really good menagerie. He has two rhinoceroses, who live in a large walled enclosure, in the centre of which is a reservoir of water, and five crocodiles live in this lake. The said crocodiles are fed with young pigs, which are turned into the enclosure ; and when, unconscious of their danger, they gó down to the water to drink, they fall into the jaws of the grim alligators.' There is however, one full grown pig now in the enclosure, who has grown up there, having survived the perils of his early youth in this dangerous place. It seems that he all at once took a fancy to the rhinoceros, and when the rhinoceroses went down to drink, he went with them, and managed to escape under their legs from the rush of the alligators. Since his first escape, he only goes to drink under the protection of the rhinoceroses, with whom he is on the most familiar terms. I was told that he looks on with perfect unconcern at the fate of the young pigs which are now sent in to feed the crocodiles, and never attempts to associate with them or warn them of their danger."

Now many may think this an unlikely story, but, barring my cousin's observation, I can assert that the pig is one of the most "talented of beasts." I have had conversations with men who have trained "learned pigs," and have learnt their secret, which I am not at liberty to divulge. Suffice it that, from what I learned from the pig-trainers, and my own observation, I can assert that the pig has a good brain, and knows how to use it to the best advantage.

When hauling the deep sea lines for "rigs" and other kinds of dog-fish, the Folkestone fishermen sometimes find they have caught a "*Kettle-maw*." I had some difficulty in finding out what this curious fish could be; it turned out to be an old friend, whose fin I had often shaken in Bond Street. Who has not seen *Lophius piscatorius*, *alias* "the sea-devil," *alias* "the wide-gap," exposed to the gaze of the Londoners in Groves's shop? Of all hideous fish, I think this is the ugliest. He looks like a huge fresh-water tadpole, and is like that wonderful hero of the nursery, one "Hoddy Toddy," of whom it is recorded that he was "all head, and no body." When the jaws of this fish are pulled asunder, his head expands into an immense mouth—a good model, as far as the gape is concerned, for the artists who make those gigantic pasteboard, open-mouthed, human faces we see at the toy-shops, and afford children sport to throw balls into. The "kettle-maw" has a natural fishing-rod fastened to his head; it is in the form of a long rod of bone, to the tip of which is attached a loose bit of skin. The joint by which this rod is fastened to the bones of the head is exceedingly curious in its anatomy. You may see a model of it if you close the forefinger and thumb of the left hand and hook the right hand round the circle thus formed. The fish, who is most happily called "the angler," conceals himself in the mud and waves, his fishing-rod above his body. Small fish are attracted by it, and, when smelling about the bit of skin that acts as a bait, are suddenly snapped up by the great mouth that all of a sudden rises up from below, and sucks them into its gape. There are several stuffed specimens of this fish in the British Museum, and the visitor will see at once how well this natural fishing-rod and gigantic mouth are adapted to act in concert.

When hooked out at sea, they are lazy brutes,

and don't struggle or try to get away. The fishermen don't attempt to sell them for eating purposes, but simply to show about the town on a tray (one was shown last year of an enormous size,) and this exhibition will often collect a pretty good lot of halfpence in the old teapot used as a money-box. Sometimes they are sent up to London to be sold as curiosities. These peripatetic exhibitions of sea monsters are profitable things "at times." An ingenious man, quite a Barnum in his way, well knowing that the knowledge of natural objects was not over rife in country places, selected from a trawl-net two large tortoises (or cuttle-fish ;) he spread them out neatly on a clean tray, and by their side he placed five other "tortoises" of graduated sizes, the last in the row being quite a little one. These were exhibited as "the father, mother, and five children," and they got him a good dinner and a skin-full of beer." A lady at Dover, not long since, bought two very curious fish; they were, the man told her, male and female.

They proved to be the upper and lower jaw of a species of skate, nicely cleaned and dried.

Cod-fish come in within three or four miles of the shore in September, and stay till April. The men look out anxiously for them to be caught in the trawl-nets, as this gives notice of their arrival. The animal from the whelk-shell is the best bait for cod; a salted sprat is also used. The whelks are caught by simply letting down open-mouthed baskets, with odds and ends of fish for bait, among the rocks. The whelks, who are horribly carnivorous, will eat anything high-flavored, and soon crowd into the baskets; they are then easily hauled up and carried home. I have heard terrible stories of these whelks devouring human bodies found floating out at sea from wrecks—one in particular, where the flesh of the face and hands of a sailor had been eaten off, besides other instances too horrible to repeat. I always shudder when I see people eating whelks in the London street stalls.

Two or three cods were caught in August last, miserable looking things, with enormous heads;

they call them "skeleton cods"—a famous name for them, as they look like ghosts of a good Dogger-bank fish, as he appears at Christmas time. These unfortunates had been living on "sea mice"—these curious creatures, that look like a large black garden slug, covered with a coat of the most beautiful iridescent hairs. Two "skeleton cods" sold for 9*d.*, to be salted down. When without their heads, they would not look so bad.

Before the cods arrive, during the summer months, when there is no wind, and the boats cannot for this reason go out trawling, the fishermen occupy their time in catching "pouters," *alias* "pouts," "bibs," "blens," "blinds," and "stink alives;" the latter elegant but appropriate name is given them because they so soon become unfit to eat after being caught. This pouter-fishing is good sport when it is a "fine catching day." The boats are anchored over the rocks, lines are sunk with leads attached, and at right angles to the lines project what are called "chopsticks;" two old umbrella iron ribs make capital "chopsticks." The bait (the thick end of a lugworm) floats free on the end of a "snood" in the tide-stream; and if there is a pouter near, he is sure to take the hook. It is a curious fact that if a pouter be taken out of the water and put in again, the moment he is released from the hook, he "gets blowed and can't swim," but floats on the surface of the water, till a "sea-maw" (gull) spies him, and "whaffles him down." When flopping about at the bottom of the boat, the pouters are of the most beautiful iridescent colors, which play about their quivering skins, and bands of a dark color are seen on their bodies. They soon assume a dull "brown paper" color, and the bands quite disappear. These fish have strings depending from their mouths, like cod, whiting, &c.; and these feelers are endowed with great sensibility, for the fish jumps about famously when they are touched, and if pinched hard, they will bring an apparently dead fish almost to life again. They doubtless act as feelers to these fish, who live in broken rocks, and may also serve as baits to the crabs—for, as I ascertained, they feed mostly upon crabs; and the process of digestion seems to turn the crab-shells of a scarlet color, as if they had been boiled.

The pouters come in inshore water in June, and go away again in January. The best bait for the latter are sand-eels. The "pouters" are frequently, by an ingenious process, "made into whiting;" and I saw, one very calm day, an old man in the harbor hard at work at this manufacture. With a sharp knife he very carefully cut out the gills and the feelers, and then sliced a great bit off the top and bottom of the fish, and nipped off the remaining fins and poked out the eyes. He then stripped off every bit of skin from the whole fish, head and all, and twisting round the tail, passed it through the holes where the eyes had been, and over a certain bone which kept it fixed there; a neat twist of the tail prevented its coming out again, and the tail-fin itself was neatly rounded off close to its stump. The fish was then put into cold fresh water to stiffen, and afterwards exposed for sale, coiled up in the most approved whiting fashion. The pouter, before being "doctored," was worth a penny—he now sells for sixpence; and I defy the whole British Association to make out his true nature when thus metamorphosed—whether whiting or pouter; but it does not much matter as far as *eating* goes, for the one is as good food as the other when nicely sent to table. By a somewhat similar process "Uncle Owl" and the "sand-rate" (ray) are made into "crimped skate" for the market at Ramsgate and Margate. There is likewise an ingenious process of "stuffing a skeleton cod," so as to render him fat looking and more saleable in the market.

The thornback family have, as the reader well knows, a fearful armament of prickles and spines upon their fins, their backs, and on their tails; and these thorny and uneatable portions, and sometimes the spines themselves, are found in abundance in Folkestone harbor. Now, hereby hangs a tale, which I published in "Household Words." It runs as follows:—

When, in the month of June, 1855, the greatest Greek scholar of the day, the Rev. Dr. Gaisford, Dean of Christchurch, Oxford, was carried to his last resting-place within the walls of the ancient cathedral over which he had presided so many years, the students of the house, clad in white surplices, preceded the remains of their venerated Dean as the procession passed along the east side of the quadrangle from the deanery to the cathedral. Great Tom had, by tolling every minute (a thing never done, except at the death of the sovereign or the dean), announced the decease ; and now a small hand-bell, carried in front of the procession by the dean's verger, and tolled every half-minute, announced that the last rites were about to take place. I was, of course, present at the funeral, and a most impressive sight it was.

The cathedral clock struck four ; the usual merry peal of bells for evening prayers was silent. I strolled towards the cathedral, and, finding a side-door open, walked in. The dull, harsh, and grating sound of the workmen filling up the grave struck heavily on my ears, as it resounded through the body of the church. The mourners were all gone; and alone, at the head of the grave, watching vacantly the busy laborers, stood the white-headed old verger; another hour, the ground would be all levelled, and the stones replaced over the master he had served faithfully so many years.

The verger informed me that the ground now opened had not been moved for 200 years, and that a Dean had not been buried within the precincts of the church for nearly 100 years. Bearing these facts in mind, I poked about among the earth which had been thrown out of the grave. I found among the brick-bats and rubbish a few broken portions of human bones, which had evidently been buried very many years; but, fastened on to one of the brick-bats, I discovered a little bone, which I at once pronounced not to be human. It was a little round bone, about the size of a large shirt-stud, from the centre of which projected a longish, tooth-like spine, the end of which still remained as sharp as a needle, and the enamel which covered it still resisted a scratch from a knife. The actual body of the bone was very light and brittle, and a simple test I applied showed that it had been under ground very many years.

The question arose, what was this bone, and how did it get to the place where it was found? It was shown to the greatest authority we have in comparative anatomy, and he immediately pronounced it (as I thought it was) to be a spine from the back of a large thornback. The creature has, fixed into the skin of his back, in a row along the back of his tail, many very sharp prickles of a tooth-like character, and covered with enamel, just like our specimen. If one of these skin-teeth be cut out from a recent fish, the stud-like knob of bone into which the spine is fixed, will be found, serving to keep this formidable weapon (for such it is) in its proper position; and dreadful blows can Mr. Thornback give with his armed tail in his battles, be they submarine, or be they in the fisherman's boat.

How did the spine of a thornback get into

Christ-church Cathedral, into ground that had not been moved for 200 years? I am indebted to the gentleman above mentioned for the idea of the following explanation, which I think will "hold water." Before the days of Henry the Eighth, the precincts, where the college now stands, were occupied by monkish buildings, where monks had many fast-days, and, on these days, were great consumers of fish. The supply of fresh-water fish, from the Thames close by, would hardly be equal to the demand. It is therefore probable that they procured salt-water fish; and a thornback is, above all fish, the most likely to have been supplied by the fishmonger.

In an old book on fishes and serpents I found unexpected evidence to prove that these fish, a hundred years ago, formed a favorite dish at the high tables of the colleges. The book was published in 1763, and the passage runs thus:— "The skate, or flaire, is remarkably large, and will sometimes weigh above one hundred pounds; but what is still more extraordinary, there was one sold by a fishmonger at Cambridge to St.

John's College, which weighed two hundred pounds, and dined 120 people. The length was forty-two inches, and the breadth thirty-one inches."

The monkish cook, like a cook of the present day, would, probably, skin and cut off the tail of the thorn-back, when he cooked him for the monks' dinner, and then he would probably throw both skin and tail, spines and all, into the rubbish-hole outside the kitchen; there they would remain till removed. And, next, when did this removal take place? A curious book ("Collectanea Curiosa") published at Oxford in 1781, tells us. In this book there is an article entitled, "Out of the journal book of the expences of all the buildings of Christ Church College, Oxon, which I had of Mr. Pore, of Blechinton."

The second item runs thus: "Spent about the femerell of the new kitchen and sundry gutters pertaining to the same, xviijd."

Further on we find, "Paid to Thomas Hewister, for carriage of earth and rubble from the fayre gate, and the new stepull to fill the ditches, on the backside of the college, clvj. loads, at a peny the load by computation, xiijs."

Again: "Paid to Mr. David Griffith, Priest, for his stipend for wages, as well for keeping of the monastery of St. Frideswide, and saying of Divine service after the suppression of the same unto the first stalling of the dean and canons in the said college, as for his labours in overseeing the workmen dayly labouring there in all by the space of thirteen months, vij£."

From this evidence it will appear that for a considerable space of time (probably about five years) many alterations were made, and much earth removed from place to place. The Cathedral of Christchurch, and, in fact, nearly all the quadrangle—as will appear by comparing their levels with that of the street outside—stand upon made ground. It is probable, therefore, that some of the earth from outside the monkish kitchen, or other rubbish-hole, was carted to form the floor of the cathedral, and with it, of course, any rubbish that happened to be there.

This, then, was the fate of our thornback's spine.

The thornback was eaten by the monks of St. Frideswide, the spine thrown away, unheeded, unregarded, to be disinterred, after the lapse of more than 300 years, at the funeral of one of the greatest and most learned men who ever presided over "The House."

One day at Folkestone, when it was blowing hard, I saw the same old man mentioned before, looking out towards the sea with the most dismal countenance. "What is the matter?" said I. "Lord, sir, it's hard times; I have not catched a 'pung' or a 'heaver' in my 'stalkers' this week; the 'man-suckers' and 'slutters' gets into them, and the congers knocks them all to pieces." My friend was a hearty old man, over eighty, and gained his living by sinking among the rocks his "stalkers," *i.e.* crab-pots, made of hops and nets, in order to catch "pung" or "heavers," that is, crabs. The "man-suckers" and "slutters" which annoyed him are cuttle fish and jelly-fish. The poor old fellow gets a scanty living by paddling about all day in his boat in fine weather; but I fear he will one day lose his life in his occupation, for he has twice been found watching his stalkers, helpless and half frozen, and has been towed into Folkestone harbor by the trawl-boats.

As his "stalkers" are always down, he has bad sport, because "the rocks is all catched up"—*anglicè* he has caught all the crabs that are about. The great even in his life was the capture of a crab, for which he got two shillings at the "Flower-de-luce," *i.e.* Fleur-de-lis public-house. He is contented now if he gets a dozen little crabs the size of a coachman's buttons, worth a halfpenny each. His costume, and solemn, wrinkled, but handsome face, were so remarkable, that I persuaded him to sit for a photograph with his old hat on. This is a wonderful hat—at least fifty years old, and without a particle of nap on it. It forms his storehouse, and in which I found that he kept hooks, lines, tows, needles, knives, bread and cheese, and, last but not least, a stout bit of knotted rope coiled up at the bottom, with which he "jackets" the boys when he finds them playing with his boat. He is full of stories of the good old smuggling times, of wrecks, and of local fishing and seafaring events, which but few men in the harbor but himself can recollect. I had many a long chat with him when he was in a good humor.

There are several kinds of crabs which he catches, viz. the "Spanish crab," a green fellow "which has oars to swim with" (the edible crab can't swim,) the "spider crab," and the "adder crab," a little fellow, capital bait for "hooking" or fishing, in deep water from a boat, for whiting and "pouters," or whiting pouts; he "heaves them all overboard" except the "pungs" or edible crabs. The "pungs" alone I have observed simulate death when brought out of the water, the other kinds run about the bottom of the boat right merrily. "A good pung 'heaves' (*i.e.* bulges out) at the tail, and is thin in the shell, showing he is full of fish and not water." He can be killed by a stab under the tail. This operation is generally advisable, because Mr.

Pung "pinches on the sly without mercy, and holds on like a bull-dog;"

> With stony gloves his hands are firmly cased;

and when he has once got hold, the only way to get him off is to snap off his claw from the body. The green crab "nips ye sharp and lets go again."

A fisherman in the harbor was holding up a large pung for sale, when he was caught by one of the claws, and in trying to get it off the other claw got hold also, and made him a prisoner by both hands, to the delight of the amphibious boys who abound in this locality. Finger nails are often destroyed by the pungs, being pinched off as though in a door.

We must not despise the poor "pung;" for he is immortalized, possibly on account of his holding powers, by being placed among the signs of the Zodiac, and the following is the best explanation I can find of his being promoted to that dignity:—"Taurus, the bull, is emblematical of culture and settlement; Gemini, the Twins, are expressive of the fertility and consequent abundance following the domestication of the ox, and of the brotherhood or union of man with man for mutual advantage and agricultural purposes; "Cancer," the Crab, considered as indicating the firmness of that brotherhood, the establishment of which is typified by Gemini."

The crab, too, in that wonderful Japan, is used as a type of distinction. Mr. Oliphant writes: "We found the commissioners strutting about the deck of the yacht in all the bravery of their resplendent costumes. Higo was literally covered with crabs, some of them large enough to be an honor to an English seaport. The dress was embroidered silk, with these crabs in raised silver, standing out in high relief."

Crabs are terrible fighters, and often lose their claws, which sometimes grow again. I got one specimen without any claws at all, and no appearance of their coming again, for the wounds were covered with barnacle shells. The explanation of this phenomenon was that "he throw'd his claws when he was young, and they never grow'd again." My father's old friend, Mr. Stowe, of Buckingham, tells me that when at Carnarvon, he saw several women sitting at stalls selling bundles of crabs' claws for a penny. On asking where the crabs were, they said, "We puts them back again into the water, and they grows again." Lobsters are terrible fellows "to shoot their claws," which detracts much from their value, and when alive and freshly caught the fishermen won't allow them to be handled for this reason. When first taken, they are "awful savage, and flies at ye like a dog."

In the London markets they are obliged to stab and kill the crabs before they put them into the boiling pot, or they would cast their claws. Lobsters seem not to do so. Their pinchers are, however, always tied together by a peculiar knot, sometimes of wire, sometimes of string; they are never kept together by means of pegs of wood driven in, as there is an Act of Parliament to forbid this practice. The lobsters die directly they are put into the boiling water, and it is cruel to put them in unless the water is boiling. There is one man in Hungerford who "boils for the market" every day, to save individual fishmongers trouble. Each lobster ought to "have twenty minutes to boil;" a crab requires a good hour if large: a deal of salt must be put in with them. Winkles take about three minutes to boil. I lately saw a man in the street putting muscles into water which did *not* boil, in order to cook them. As the poor things went in there was a peculiar hissing sound, and a scum on the top of the water. I remonstrated with him on his cruelty. "That's nothing, sir; it's only the things a-fretting themselves," was the answer. I asked him if he would not "fret himself" if he was gradually boiled to death? It was a new idea that had never struck him, and he promised to use boiling water for the future.

It is curious to remark the difference in structure of the lungs of the fish, and the lobster and crab. They both live in the water, but yet, if taken out of the water, the fish dies at once, the lobster will live some hours.

May not the explanation of the structure of the lung which breathes as well in water as in the air be as follows?—The fish are supposed always to live in the water: those which live in deep water never run the chance of being left high and dry; those which live along shore, as the fisherman well knows, run in and out with the tide, as their instinct prompts them to do, in order to avoid this catastrophe. If by chance these fish are left by the receding tide, they soon die; crabs, lobsters, and cuttle-fish, on the contrary, live among the rocks; *they* can't run in and out with the tide, and they therefore often get left high and dry on the rocks, or on a fishmonger's slab, which is the same thing in effect. Nature, in her wonderful wisdom, anticipates all this (not the fishmonger's shop,) and gives them a breathing apparatus, by which they are enabled to keep alive till the tide comes up again and covers the rocks, enabling them to refill their breathing sacks. A lobster, I believe, if left on the rocks, never goes back into the water of his own accord; he waits till it comes to him. I say this, because when the landslip took place near Lyme-Regis, in Dorsetshire, a great portion of the bottom of the sea was forced up by the weight of the landslip on the margin of the shore. On this suddenly elevated bit of ground there happened to be several lobsters, who, doubtless, thought the low tide had taken place, with uncommon celerity, and that it would return again. Anyhow the brutes, obedient to their instinct (I dare not call them stupid), waited for the tide to come up and cover them. Of course it never did come up again; they remained in their places and died there, although the water was in many instances only a few feet from their noses, and they had not the sense to tumble into it and save their lives.

The crab's broad back affords a capital habitation for minor sea creatures. Sailors have observed the fact of shells sticking to crabs and other animals which live in the sea. Hence, an old patriarch sailor who has long been at sea often goes by the nickname of "Old Shellback." I have got a crab with a regular crop of young oysters on his back, and another whose shell had been pierced by some boring animal; in order to mend the hole, the crab had secreted a round nob of shell *on the inside*, and this looked very like a pearl. In the Museum of the Royal College of Surgeons is a very remarkable specimen, showing the same process of healing a wound of the shell in the true tortoise-shell turtle. The creature had been wounded in the back, probably by a spear such as is used to catch them when floating on the water; it had escaped with a hole in its shell, which nature has beautifully mended with a round nob of true pure tortoise-shell. This nob is about the size of a cricket-ball, and like the cricket-ball presents a series of laminæ on a section being made. It is a unique and most curious example of nature's mode of healing under difficulties.

Every day hundreds of little soles (slippers as they are called,) young turbots, flounders, &c., too small for cooking, are brought in by the trawl-nets and thrown away as useless. Twenty or thirty little turbots have been thus sacrificed in one day. Sometimes they are hung up in long festoons to dry, like haddocks. This wholesale slaughter in time will tell on the number of fish in the trawling ground, which is but a limited space after all. The fishermen told me they were aware of this, but could not remedy it, because everybody would not agree to use nets with smaller "shale" or mesh, and when the trawls were hauled aboard the small fish were found for the most part to be dead from being dragged so long along in the net with the other fish. I have seen a bunch of little dabs and flounders, &c., worth fourpence or sixpence for the lot to a London costermonger, tied up in a stalker to catch a halfpenny crab.

As regards the above passage, the editor of "The Field" for October 15, 1850, remarks, under the "Angling" column :—

"Mr. Buckland, in his last paper of 'Sea-side Gatherings,' touches upon a highly important subject when he deplores the destruction of small flat-fish in trawls, and says that it will tell in time on the trawling-grounds. The fish, however, destroyed in the trawl-nets do not amount to a tithe of what are destroyed in the shrimp-nets. We have seen hundreds of small soles, plaice, turbots and dabs not two inches long, cast with abominable recklessness dead upon the shore by shrimpers, when it would have been even less trouble to return them to the water. Idle boys, too, yearly destroy thousands of these small fry by way of amusement. That these practices have a most injurious effect upon the fisheries there cannot be a shadow of a doubt. We have repeatedly urged upon the public, and shall do so upon every opportunity, the extreme importance to Great Britain of her coast fisheries. It is to the hardy race of mariners bred in the following of the occupation of fishermen that she owed her supremacy of the seas; and if she is to retain that supremacy, she must, to a great extent, rely upon her fisheries as a school for experienced and adventurous seamen. Our fathers well knew the importance of this matter, and framed wise laws for the control and management of their fisheries—in many instances fostering them to their utmost power by means of bounties and rewards. The size of every fish, and under which no fish could legally be taken, was laid down in the 1 Geo. I. That Act, as far as we are aware, has not been repealed; and it provides, first, that save for herrings, pilchards, sprats, or lavinian, no net shall be used of less than three and a half inch mesh under a penalty of 20*l*. And, further, that a fine of 20*s*. be imposed for every offence in the case of a person bringing to shore or offering for sale, &c. &c. any turbot of less than 16 inches in length, brill 14 inches, codling 12 inches, whiting 6 inches, bass and mullet 12 inches, sole 8 inches, plaice or dabs 8 inches, and flounders 7 inches.

These matters have, since the date of this Act, been placed under the charge of the Coastguard, and the Coastguard have lately received orders to enforce the fishery laws strictly. A week or two since we gave an instance in which these laws had been strictly enforced in consequence of this order; and we hope that the lead thus aken will be followed up, and these mischievous and ruinous practices summarily put a stop to."

I was told of a great curiosity, a most wonderful new bait to put into the lobster-pots; I found it out, and ascertained it to be a common phial physic bottle, silvered in the inside like a looking-glass. It was given by a gentleman to Smith, sen., one of the patriarch fishermen in the harbor, who showed it to me. His theory was, that "the lobsters comes to see themselves in the glass;" but I doubt whether a lobster (even though he has "a lady in his head") has vanity enough to use a looking-glass. It must be the glistening of this bright object that attracts them, because, when bait is scarce, a "bunch of oyster-shells with the dark part scraped away" will sometimes catch them; "but they gets out again pretty quick if they don't find nothing to eat." There are plenty of fine prawns about the rocks, and there are two modes of fishing for them; first, with a common shrimping-net pushed along the sand; and, second, the "lock-nets," which are simply a large form of the round nets used to catch freshwater crayfish. A bunch of fish (with the dark skins taken off) is tied on to "the bridle," a string that goes across its centre, and a long line with corks attached; the corks are covered with white linen, that they may be seen in the dark. The "prawner" has a stick some seven feet long, with notches at the top; as the tide goes down he puts his net among the rocks, and leaving it a few minutes hauls it up on the stick with a jerk, as the prawn is an active fellow and soon jumps out. The best bait for prawns is "a fresh sheep's head." It is no use fishing for prawns till the sun is down, and then, and not till then, the prawns begin to feed. It is a curious fact that the prawns are most numerous "just as the visitors go away;" but the season has, in my opinion, more to do with this than the visitors. The crabs —the spiders, the Spanish, the soldiers, and the adders—are a great nuisance in the lock-net fishing; they get to the bait and are hauled up by the score, pinching the poor man's fingers terribly when he tries to get the prawns out from among them; for it must be recollected that it is pitch dark, and the contents of the net not easily seen. The prawns themselves make wounds on the fingers with their sharp and projecting "spears." The hands of one of the prawn boys was just as if it had been "crimped" all over from this cause. It is dangerous work fishing in pitchy darkness among the slippery rocks, "where you may knock yourself all to pieces in a minute," the tide rolling in fast all the time; and the following was the anathema of my companion, a poor man who had been hauling up red crabs, and not prawns, from nine to eleven that night; "D—— them crabs, the rascals, they hinder me of many a shilling and rob my young ones of many a slice of bread and butter; them fellows won't let ye have a prawn if there is one in the net. I will smash some of ye, anyhow," said he, as he dashed crab after crab on the rocks. However, I consoled him somewhat by buying a large bag full of his crabs, and served them out by boiling them for supper but they were very bad eating.

In the islands of the Pacific Ocean there lives a frugiverous crab, which "climb a species of palm (*Pandanus oderatissimus*), and eat a small kind of cocoa-nut that grows thereon; they live in holes at the roots of the trees, and are a favorite food of the natives." The name of this crab is *Birgus Latro*, or "the Robber." Now, the natives aforesaid cannot use "stalkers" or "lock-nets" to catch the poachers, but they have a most ingenious and simple trap instead. A good way up the stem of the palm, they tie round a thick bit of rope or matting; the crab comes down the tree tail foremost, and when he gets to the rope, he thinks he has got to the ground at the bottom of the tree; he lets go his hold, and falling to the ground, gets either killed or else so injured that he is easily caught.

Both lobsters, crabs, and prawns cast their armor when they get too big for the one they occupy. When a lobster is being eaten at table, a skin will be found under the shell, and it is this skin which will harden into a shell when the old one is cast off. I was lucky enough to obtain from a shrimping boy a lobster who had just cast his coat. The new one was soft to the touch, and like moistened parchment. It is a rarity to catch these "soft lobsters," as the creature knows that before his armor gets hard he has no mode of defence, and hides himself away in a snug corner; but my friend did not calculate upon the shrimping boys, who espied him under a rock at low spring-tide, and made an easy capture of him. I obtained last year from Mr. Townsend, fishmonger, of Hungerford Market, a fine specimen (now in the Royal College of Surgeons) of a lobster with his old coat off his back, showing the new one hardening underneath.

Mr. Townsend tells me that there are three kinds of lobsters sold in the London markets; viz. the English, which "boils heavy," and is of a white color under the claws and the tail; the Norway lobster, which is much lighter after boiling, and is principally used for making sauce; and, lastly, the Norway lobster (a coarser kind), which has a very hard shell, and is generally covered with little round white things, like hard chalk (i. e. *serpulæ*).

The light and delicate cast skins of prawns and shrimps I saw frequently floating about in the harbor, and when placed in a glass of clear water, they look like the bodiless ghosts of their former tenants.

Lobsters are very coarse feeders, and can be caught with a stinking bait; not so with "pungs" —the fresher the bait, the better chance of sport. The old man came down daily to the fishing boats, and collected the heads. entrails, &c. of the fish, which he used for bait, tying them up in a great bunch in the centre of his "stalkers."

In the "stalkers," and in the shrimp nets, fishermen sometimes catch what they call "Sea Snakes," more commonly known by the name of "Pipe Fish." This curious fish belongs to the order *Syngnathidæ*—so called because "the jaws are united," so that the mouth forms a long cylindrical tube. Now, to this family belongs the Hippocampus, or Sea Horse, which has occasion ally been taken on our own southern shores, and also at the islands of Guernsey and Jersey. A pair of these curious and pretty little creatures were lately exhibited alive in the Regent's Park Gardens. They were brought from the mouth of the Tagus, and presented by a Portuguese gentleman. Two were offered for sale, not long ago, alive, in Hungerford Market, for a guinea each. If there were more demand for them, they would doubtless be more frequently seen in our vivaria than they are at present. The Regent's Park specimens were not much larger than a good-sized sprat. The head is wonderfully like that of a horse; the eyes are exceedingly bright and prominent, and they have the power of moving each independently, like a chamelion. Their gill-flaps cannot be seen to move up and down; they are fixed down to the adjoining skin, as we see in the eel family. The actual gills, moreover, are not comb-shaped, like those of other fish, but are disposed *in tufts*, about the bones which support them. On each side of the head there appeared to be ears erect and expressive, like those of a horse when in a listening attitude. These are not ears, but tiny fins, which, when the owner begins to move, vibrate with astonishing velocity. There is also a fin on the back; and when this is put in motion, it vibrates from end to end in serpentine waves, like a bit of rope laid on the ground and shaken sharply. When in motion, it is an exceedingly pretty object. The body is terminated by a long and gradually pointing tail, which is of the greatest use to the owner; for by bending and uncoiling it, he is enabled to assist his progress in swimming: he generally carries it coiled upwards in a graceful form. But the Hippocampus is a lazy fellow, and had rather sit still than move about; and here the tail is of the greatest service, for he twines it round bits of sea-weed, projecting rocks, &c., and remains, as it were, anchored, while he sways his body backwards and forwards in search of food. It is doubtful what substance forms his food. Both the upper and lower jaws are connected together along their whole length, leaving only a very little mouth at the end, exactly like the mouth of the ant-eater. I am inclined to believe they live on infusoria and very minute sea insects, which they find abundantly at the sides and bottom of the rocky pools which they naturally inhabit. I watched one of them come and anchor himself near to a bit of sand in the vivarium, and then, bending his head downwards on to it, peck about it exactly like a hen scrutinising a dust heap. He perpetually took up, with a sucking action of the jaws, grains of sand into his mouth, and spat them out again, as if trying whether they were good for food or not; and I fear, poor fellow, he did not get much for his trouble. If it were possible to procure some of those little shrimp-like things we see hopping about the sea-weed on the shore, and which they call "skipjacks," I am convinced the prisoners would eat them with eagerness.

The Hippocami are admirably adapted for living among rocks. They have no scales; but their heads, bodies, and tails are defended by a casing of bony armor, of a pretty and complicated pattern, so that they shall not get hurt and bruised as they wander about in their rocky feeding-grounds. When viewed under certain lights, their armor, or the interspaces between its plates, seems to be iridescent, like the beautiful blue and green colors of a mackerel. This is not an exception to the rule, for most fish that live down in the dark recesses of the rocks, such as the rock-fish, whiting, and poult, display exceedingly beautiful colors on their skins when first taken out of the water. They are, moreover, capital climbers, and manage by means of projections from their armor, by hooking on with their heads, and by grasping with their prehensile tail, to scramble up and about rocks which, to the Hippocami, must be comparatively as high as Dover cliffs to ourselves. As the kangaroo and opossum are marsupial, and carry their young in a bag or pouch in the abdomen, so, strange to say, even among fish we have an example of a marsupial creature. In the *male* Hippocami, contrary to what we should expect to be the case, we find that the bony armor in front of the abdomen is converted into a sort of leatherlike bag; and into this the young escape at the appearance of danger, and they would rather swim into this place of concealment (as has been observed) than to any neighboring bit of sea-weed or crevice in the rocks. We have several preparations at the college of Surgeons to show this very remarkable part of their economy.

I regret that both these interesting little animals are dead. I wished I had been enabled to have supplied them with plenty of "skip-jacks," which I am sure they would have eaten.

As the steamer from Folkestone paddles along towards Boulogne she crosses over two sandbanks that lie about mid-channel. These are doubtless two sub-aqueous chalk hills, the summits of which have been covered with sand by the action of the tides; between these hills deep water (*i. e.* a valley) is found. These banks are called by the fishermen the "Warne," or "Werne," and the "Ridge." Congers, dog-fish, basse, &c. form the game in these submarine preserves, and to catch them the fisherman "shoots" not his gun, but his "long line." Besides the Warne and the Ridge there is another naturally formed longer preserve "off the Ness," *i. e.* "Dungeness Head," about eleven miles by sea S.E. of Folkestone. Extending several miles from Dungeness towards Hythe there is a long dreary beach, composed entirely of loose shingle. It is exceedingly difficult to walk for any long time on this shingle; it is just loose enough to admit the foot and ankle at every step, and in a few minutes the pedestrian becomes exhausted, and can hardly proceed on his way, or, in fisherman's language, "the stones would pull your legs out of the sockets of your body." Business compels these poor fellows to walk on the shingle, and necessity being the mother of invention, they have contrived a species of shoe or patten, on the principle of the Esquimaux shoe, or the mud-shoes of the wild-duck shooter. They consist of light boards, about the length of an Esquimaux shoe, upon which is a strap to insert the foot. On these they are enabled to traverse the shingle without sinking in. The local name for these ingenious contrivances is "Bexters;" the derivation of this word is a mystery to me, unless a man of this name invented them—thus becoming immortalized, as the memory of our great surgeon, Abernethy, is daily called to mind when we ask for sixpennyworth of Abernethy biscuits. The water just at the Ness is exceedingly deep; even close up to the cliff there is from twenty to thirty fathom water. Down among the rocks at the bottom of this hole or natural pond the "great" congers hold nightly revel. Their mid-channel brethren are not nearly so large. The first sight I got of the congers at Folkestone was in a "lugsail boat," which ran in at high tide in company with a fleet of trawl-boats, laden with congers, or rather marine boa-constrictors. When the boat stranded, the men threw them out on the shore one by one, and there they lay, just able to wriggle, and to gasp with their formidable mouths. Some of them were of a pale white color, but the majority were sprinkled here and there with nut-brown markings, making them look much more snake-like than when hung up in Billingsgate. The crowd gathered round. "Who will buy this parcel of congers?" said the fisherman, picking up a stone, and standing with the congers all placed before him. The biddings went on fast. £1 5*s.* was bid for the lot, in number twenty-one, large and small; down went the stone, and the purchaser hastily pitched the great brutes into "kittens," as they call the fish-baskets, and in twenty minutes the congers passed me in the open luggage-van of the train on their road to London. These were "Ness congers," and are caught on a "long line." The lines laid for them are seventy-five fathoms long, and on each line are attached at right angles other smaller lines, viz. the "snoods," twenty-three snoods to each line, each snood nine feet long. The hooks are nearly as large as the hook of a roasting-jack; they are made of exceedingly tough wire, that will bend to any amount, *but never break.* Strange to say, these hooks can't be bought in England; they are all of French make, and cost about 2*s.* 6*d.* a hundred. Each hook is fastened by a simple but very firm knot to the "snood," by means of a tough bit of cord called the "beckett"—a wise provision, for the congers "saw themselves on the rocks, and often burst the line;" and when they are first hauled up there is no getting the hook out of their mouth or throat, as the case may be. The "beckett" is therefore cut, and the conger slips into the bottom of the boat, to die at his leisure The conger "often gorges the hook," and it gets fast a long way down from his mouth. When the fisherman arrives on shore he has a summary way of getting it out; he takes hold of the "beckett," and smacks the conger as you would smack a hunting-whip; the hook soon straightens and comes out; a tap with a hammer makes it fit for service again. The line complete is worth between 7*s.* and 8*s.*, and as many as nineteen lines are sometimes laid down in a night. Thirty congers is a good haul; formerly eighty or ninety were caught, but then "they were of no use, because there was no railway, and therefore no ready sale." At present the Billingsgate salesmen will take any number, and the congers, in consequence, are "more gone after." Those that are not sold for food in London are made into isinglass. One Fagg, a fisherman, told me that three years ago "the more stinking his congers were, the more money they fetched in London. He never could understand this, but it was all the better for him, as congers don't keep over well in hot weather. They lives as long again in cold weather as they does in hot."

When the long line is hauled the hooked congers sometimes (if they are big ones) "yawls on the lines;" sometimes "they comes up quietly;" at other times "they hangs like a log of wood;" but they nearly always "makes a curl with their tail, so as to hang back in the water;" when in the boat they try to get out by "clinging their tails over the side." (I have observed this same fact with fresh-water eels.) To get them into the boat the fishermen have an enormous "heaf" or "prule," *i.e.* a gaff-hook. When in the boat "the congers are terrible things to bite sure-ly." To kill them, "it's no use knocking them on their great heads, no more than a great bull; just hit 'em a sharp smack on the belly, and that turns 'em up directly, because all their blood lays there."

One of the fishermen, George Smith, caught a "whacker" last year. When he hauled the line he thought "he had got hold of a wreck," but he managed to pull him up gently to the side of the boat, and whip the "heaf" into him. He "kicked up Mag's diversion" in the boat, and nearly got out again; so Smith tied him by the "beckett" to the thwart of the boat with a new French whiting line, which Master Conger broke three times. Smith himself, who is a very powerful man, tried afterwards to break the same whiting line, but could not. This conger measured eight feet within an inch, and was twenty-six inches in girth at his fins. He was sent off to London directly in a "kitten," or fish-hamper, all to himself. I myself measured a conger on the beach twenty-four inches girth and five feet seven inches long. I had no time or means of weighing him, which I regret. He had two soles and a flounder inside him. Congers are "the daintiest fish out. They won't look at a bait that is the least tainted; only lobsters and bass eat stinking fish." The surest way to catch them is to "trawl for the bait (small flounders, dabs, plaice, &c.) as you go along to the Ness." On these expeditions the boat must be out all night, shooting the lines at sundown, and hauling them to sunrise, or as soon after as the tide will serve.

One day a French fishing-boat came into Folkestone from Portelle, near Boulogne, the weather being too rough for them; they had a few small congers on board, and I observed that they had been baiting their "snoods" with the arms of cuttle fish cut into bits. They called all Englishmen "John," and when I went to talk to them a red-nightcapped fellow held up a dog-fish, and said in broken English, "Vil you buy a dog, John?" I did not buy the dog, but I got a conger's head for threepence, and was told by Mr. Warman, the fishmonger, soon afterwards, that I was liable to £50 for buying in the harbor fish of a Frenchman. The Frenchmen turned out of their boats in the afternoon, and boiled the conger's body on the beach, putting sundry odd scraps of fish into the pot as well. As a bystander said, "There's very little fish as them chaps heaves away; they eats a'most anything.' Poor fellows, they were very poor and very hungry, and amazed at the small quantity of *eau de vie* they got for the money *pour boire* I gave them.

There is another more curious way of fishing than by the long-lines, at present practised by not a few persons. What does the reader think of an iron hammer as a bait? To lay the foundation of some new works in the island of Alderney divers are employed; these men, enclosed in their India-rubber armor, see strange sights at the bottom of the ocean. The fish, and no wonder either, at first are alarmed at the unwonted apparition, with its huge glass goggle eyes; but, soon recovering confidence, approach to satisfy themselves of the real nature of the intruder. The monster raises his hammer which he has brought with him to quarry the rocks; the curious fish come up and inspect it; while doing so they receive a sudden knock on the head which stuns them; and, when they recover their senses, they find a bit of string through their gills, and themselves prisoners tied fast to the India-ruber monster.

On one occasion, a diver had a fight under water with one of the rightful inhabitants of the rocks, which he was so unceremoniously breaking up. A huge conger eel suddenly started from its favorite hole, and furiously attacked the destroyer of his home. A short but severe combat, between the eel and the man, ensued; but a well-directed blow of the hammer soon settled the question against the eel.

In the first series of "Curiosities of Natural History" I have recorded a curious fact, communicated to me by Dr. Turner, of Hastings, relative to the action of intensely cold weather upon the air-bladders of the congers. "In January, 1855, thousands of congers were found floating upon the water. They could progress readily in any direction, but could not descend, and consequently fell an easy prey to the boatmen. In this manner no less than *eighty tons* were captured," &c. &c. The Folkestone fishermen, I found on inquiry, also took an enormous quantity of congers at this time. One man brought home 800 in his boat. The same cause acted on the Folkestone as upon the Hastings congers, viz. the action of the frost caused the air in their swimming-bladders to expand so much that the ordinary muscles could not expel it at will.

The Folkestone fishermen explained this phenomenon thus: "You see, sir, the congers comes up to the top of a frosty night to *look at the moon*, gets nipped by the cold, and can't get down again." Barring the moon part of the story, their theory is correct.

Mr friend Mr. Roberts, of Dover (a great observer,) has kindly sent me the following note on this subject: "After a sharp snow-storm I saw picked up behind the cob at Lyme Regis, Dorset, several hundred red rock, or sea perch, called locally 'conners,' small 'blins,' *i. e.* whiting pouts, and a conger eel or so. The people come to pick up the fish *blinded by the snow* on Christ mas-day. They were lying stranded among the sea-weed. There may have been five hundred fish at least of eatable size, and about three hundred smaller ones. Three years ago a respectable man named Wood told me that he had made a good meal upon a present of conger eel, made him by the captain of a trading vessel. The vessel was bound to London in some very severe weather in January. Near the mouth of the Thames the frost and snow had so affected the congers that the crew of the trader had speared and salted several magnificent fellows. A vessel had gone to catch the blind congers, and had sold one hundred and fifty tons weight of them in Billingsgate-market. The fact of the effect produced upon the fish along shore I can attest from very frequent observation.

I was not able to go out to the Ness to fish for congers, but neverless I got my sea fishing-tackle ready, and fished from off the end of the harbor pier, on the chance of catching one of the small congers that are sometimes found there ; although fishing in the sea, the first fish I caught was a regular freshwater fish, viz. an eel, with not a bit of the conger about him. Though brackish, the water is quite salt to the taste, and regular salt-water fish, such as the plaice, bass, whiting poults, horse mackarel, &c. live in it, for I have caught specimens of all. There are plenty of eels in the mud of the harbor, and some of them are occasionally taken of a large size. They must breed there, for the simple reason that there is no exit for them except towards the sea, save except a sewer, the hatch of which is occasionally opened to flush off the water; the stones over which this sewer-water runs are covered with a green weed, and in this green weed I observed boys catching little eels about as big as a quill toothpick, and matching in appearance the young eels from the Thames, which are sold in Hungerford market just at this time of the year for vivaria specimens. I am almost certain these eels were bred in the harbor; the migrating instinct was very strong in them, for they were all crowding up to the hatchway, beyond which of course, they could not proceed. An old man who has been at Folkestone all his life, tells me that he is sure the eels breed in the harbor, and that they can be caught there all the year round.

I tried my luck fishing in the military canal near Hythe, but as they had let the salt water in, they had killed all the jack, perch, and roach, and I caught nothing but a few eels. I believe these eels breed in the canal, because if they went down to the sea they would find no mud, but only a bank of loose and rolling shingle, a most unfavorable place for them to deposit their ova. Again, there is every reason to believe they breed in ponds whence there is no exit, as has been asserted by numerous observers.

The number of eels consumed by Londoners in the form of stewed eels and eel-pies is incredible. The great proportion of these fish come from Holland, and the visitor to Billingsgate-market will always be sure to see some Dutch eel-boats at anchor in the river just off the market. Anxious to obtain information from head-quarters, I have made an excursion to the market on purpose to examine the internal economy of these boats, and learn how they carry the eels, &c. The fiery sun had caused the market to be nearly deserted, and the fish, shrimps, &c. exposed for sale emitted odors that were none of the sweetest. Getting on board a wherry, I was soon alongside the nearest Dutch "skoot," as this kind of vessel is called by the men about the market; on it were three men, two indulging in a nap, the third smoking one of the longest pipes I ever saw. Making a polite bow, I asked leave to come on board, which was immediately granted by the master—he of the pipe—one of the tallest, best-built, grey-eyed Dutchmen I ever saw.

Letting myself drop down through a hatchway, I was obliged to stoop low to walk along under the deck, through an apartment which ran from the dwelling-cabin in the stern to the bows of the ship. On one side I saw four leaden tanks, in which I was told the eels were carried. I could not understand how this could be (as they were not big enough) till the Dutchman explained that under the floor whereon we were standing was a false bottom, running nearly the length and breadth of the ship, and that these four tanks communicated with a large chamber or reservoir in which the eels were kept. Holes were bored in the sides and bottom of this reservoir to admit the water to the eels. The sea-water enters freely through these holes, but the eels don't mind this, and live very comfortably, even though they are surrounded by salt water during their passage across the Channel, which occupies from four days to a week. It is far otherwise with the Thames water. I asked if there were any eels on board at that moment. He smiled with astonishment. "Why," said he, "they would die if put into Thames water up here by London Bridge in ten minutes; the water has been getting worse and worse, and the eel-boats have been obliged to anchor lower and lower down the river. It is many years now since eels would live at London Bridge. The boats have gone first to Erith, then Greenhithe, and now they cannot come up further than Gravesend without killing the fish." The boats arriving from Holland stop at Gravesend. There are always two or more boats moored off Billingsgate, and every morning eels are sent up in flat boxes to these floating shops from Gravesend, to be sold to all comers from their decks by the pound.

Here, then, we have additional evidence of the excessive impurity of our noble river—even eels, that will live almost anywhere, can't exist in its waters, except at a great distance from the mouths of drains and other abominations, which pour into it a poison reminding us of the plagues of Egypt. In order to get the eels out of this false deck, the Dutchman showed me a net with small meshes, mounted on a pole, of a circular shape, and with this he could, from one or other of the tanks, get every eel out of the false bottom in the vessel. The ship (her name was *Cornelia*) sailed from Texel. The Dutch fishermen brought the eels on board for sale. Sometimes there was a cargo ready for them on arrival, sometimes they had to wait till a sufficient quantity was collected. And how much is a cargo? Only fancy; *six tons!*—(Query, how many eels go to a ton?)—of slimy, crawling, slippery creatures, all alive, and packed one upon the other in the false deck of the skoot. When the water on either side of the channel is low, and there is danger of the skoot grounding, Mr. Dutchman takes out from between decks two enormous wooden chests, exactly like large coffins, full of little holes; he fills these with eels, and tows them along behind the vessel, and so lightens her not a little. He never feeds the eels, and they will live at Gravesend for a week or ten days; he never saw eels larger than five or six pounds; they are very scarce this time of year.

The fishmongers, costermongers, and others come to buy on the deck of the *Cornelia*, and later in the day we see the eels hawked about the streets. I have observed that the venders, under the pretence of not being able to handle the slippery eels, cover them with sand, but they *don't wash the sand off when they weigh them*;

the customer, therefore, as Professor Quekett wisely remarks, when he buys eels thus treated buys wet sand as well, at about the rate of 4*d*. a pound—*caveat emptor*.

The last week in August, when fishing for "pouters," eels, coal whiting, and "bull routs," off the horn, at Folkestone, one of the largest fishing-boats in the harbor sailed away to the northward—the contrary direction to the usual fishing-ground. I learnt that she was gone to Yarmouth for herrings. In a few days she returned, and on going on board saw her hold quite full of these useful fish, salted, it is true, but yet glistening as though their scales had been electrotyped with silver. Gangs of men carried them away immediately to the "herring hangs," or, as they are called elsewhere, "herring dees." Of course I followed them, anxious to learn the process of "red-ding," or "bloating" herrings. The "hang" consisted of a very lofty square brick room, which can be made perfectly smoke-tight; stout beams traverse it from top to bottom, supporting a ladder-like frame-work, the bars of which are called the "rangers;" upon these "rangers" are placed the "spits," *i. e.* bits of wood about the length and thickness of a common walking-stick—these are best made of pine-wood ; upon them the herrings are hung up.

When the herrings are brought from the boat, they are washed as soon as possible in large tubs, by poor women, who are only too glad to earn a few shillings at this easy work. Six shillings is paid "per last" (N.B. "a last" is 10,000, with 100 given in for broken fish) for washing and spitting. They tell me that on these occasions the female tongues work as merrily as the female hands. This chance of "big talk," as the Mandan Indians call it, is never thrown away.

Active boys are employed to climb up with the herring spits; and this is by no means easy work, as some of the hangs are over forty feet high, the wood slippery, and they often have to work in a cloud of smoke.

When all the fish are hung up, the doors and windows are closed, and a fire made on the brick floor of logs of *oak wood* and *oak sawdust*, and there they remain enveloped in a dense atmosphere of the most pungent smoke (it makes tears come into the eyes) for several hours, till they attain the peculiar flavor which every one knows.

The reader has doubtless remarked that the red herrings in the shops, or those which fill the baskets of the loud-voiced costermonger, who at early morn disturbs the Londoner with sleep-dispelling cries of "yar-Mouth, yar-Mouth," have their gills *on one side* distended, and their mouths wide open and gaping. This is the hole made by "the spit," when the fish was hung up to dry, and which remains open after it was taken out. When, indeed, the retailer comes to the hang to buy, he brings his own bundle of spits, and the row of fish are slid on to them from off the spits on which they were dried.

When there is no fish brought into Folkestone, the bare legged boys buy a spit or two on speculation, to hawk about the streets.

It naturally happens that among 10,000 herrings or more, there are some which are "gill-broken," and therefore cannot be hung up by the heads; they are therefore tied on to the spits by the tail, and have on this account received the appropriate name of "tie-tails," or "scraps," —the London costermongers call them "plucks." Though they are just as good eating as the others, they fetch less money; and when I was in the hang, a tiny child came in and addressed the burly owner thus:—"Please, sir, mother wants a farthing's worth of tie-tails for her tea." She got two or three, and some broken "scraps" into the bargain.

There is great art in salting the herrings to the right pitch; and they should be in salt forty-eight hours if wanted to keep for the winter,—if for quick sale, ten hours in salt is sufficient. They ought to have one ton of salt to each last. "If there is not enough salt, they won't hang in the smoking, and when the door is opened they are found to be dropping down one by one, like ripe plums, leaving their heads up aloft on the spits." The way to know whether they are properly salted, is to take the fish by the head and tail, and give it a gentle pull; if it feels moist and "lissom," it is not properly cured; if, however, the bones can be felt to crack smartly, make your bidding for the lot.

The men who go to Yarmouth salt the fish as they buy them. Those at the bottom of the boat are the best, as, being the first lot put in, they get the drainings of those above them. The experienced eye will tell from the appearance of their salting the different catches put in the boat at different times. If they stick together, or have begun to ferment, they are all spoilt: so that much judgment is required to buy herrings, as it is no joke to buy several lasts which "go off" before they are fit for sale. In this condition they are not fit even for manure. The experiment of digging them into his garden was tried by one of the men, but he found his vegetables grown on the soil tasted "fishy"—and a fishy potato is not the best of eating.

When the shoals of herrings come off Folkestone (and they ought to be there now), enormous quantities are caught by the fishermen, to whom this wonderful migration is as good as a dividend at the bank. Mr. Warman, fishmonger, told me that he bought "fifty-three lasts" of herrings the season of 1858. I saw several hundreds still in the "hang." It must have been a curious sight to see 583,000 (or more than half a million) herrings all on the spits being "red-ded." At Yarmouth there must be very many more than these cured by herring-dealers, but I can write only of what I saw.

Each Folkestone herring-boat carries a "fleet" of nets, and sixty nets make a "fleet." A net, which is about thirty yards long, is made four "rans" deep, and there are sixty meshes to a "ran." The nets are made in two or more breadths joined together, each of which is one fathom in width, and is termed a "ran." To supply the wear and tear, the lowest "ran" is removed after every season, and a new top one added, so that the net is thus constantly kept in repair. I learnt, but lately, from a gentleman, (Mr. Jefford, surgeon, whom I met on the top of the Bridport coach), that they are now making nets of cotton. At Bridport, Dorset, there is a large manufacture of nets, &c. from Russian hemp, and Russian vessels are often in the harbor. Cotton nets will not do for *deep-sea* fishing; they are not strong enough. Cotton nets catch more fish than hemp nets, when used near the surface. An experiment was tried—I believe, in Cornwall—by placing alternately cotton and hemp nets. The cotton nets caught most fish; both kinds of nets were fishing under exactly the same circumstances. Nets are very expensive things, and soon get injured by the sea water. I observed that all nets at Folkestone, and even the men's frocks, were colored of a deep mahogany color. This I ascertained was caused by their having been boiled in a preserving or tanning material. Experience has taught seagoing folks both here and elsewhere that "cutch," *i. e.* "catechu," is best fitted for this purpose. This catechu was formerly called *Terra japonica*, as it was thought to be of mineral origin; it is obtained from the wood of the

"Acacia Catechu," a tree, native of the mountainous parts of Hindostan; but we luckily need not go so far as that to get it, as it is to be bought of Mr. Henry Barnes, drysalter, 38, Long Acre, W.C., at sixpence (retail) a pound.

I mention this catechu, because I think we fresh-water anglers may use it to great advantage; it may be doubtless known to many of us, but I have never seen lines and nets tanned with it in the London tackle-makers' shops. All that is required is to boil the landing-nets, lines, and other items of the angler's kit, in a strong solution of it for an hour or so, and then let it dry. I have no doubt that sixpennyworth of catechu, and a little trouble, would save many shillings, if not pounds, in the wear and tear of fresh-water as well as salt-water tackle.

It appears, from a most interesting book, entitled "Ancient Egypt, her Testimony to the Truth of the Bible," by William Osburn, Jun., 1846, that the ancient Egyptians used nets for fishing made of leather, afterwards greased, for we read:—

"The four men represented (in the drawing) are engaged in a process for the manufacture of leather, which, as the inscription informs us, fitted it to make sandals and nets for fishing. The 'inverted boat' signifies 'nets' elsewhere; the hieroglyphic group is the 'Coptic fish;' the link represents the mesh of the net, and determines the whole group, which reads 'fishing-nets.' The use of leathern wine bottles by the ancients is perfectly familiar to every reader of the New Testament (see Matt. ix. 17.) It was not so generally known that these skin vessels were of elegant shapes, and designed for ornament as well as mere utility. These pictures (there are many illustrating the manufacture of leather among ancient Egyptians) abundantly prove the existence of the manufacture as a mechanical art among the Egyptians at the time of the Exodus, and therefore that the Israelites, who had been captives there, would be well able to perform this portion of the service of the Tabernacle." These groups of Egyptian workmen strongly remind one of the old song—

"Oh that the man in heaven may dwell
Who first invented the leathern bottle."

So important is the tanning-net process with catechu to the Folkestone fishermen, that a charity has actually been established to tan the nets for the poor fellows.

About 1674 A.D. Sir Eliab Harvey, brother to the immortal Harvey who discovered the circulation of the blood (1628), both founded the Folkestone Grammar School and provided a tan-house, for the support of which the fishermen mere to pay eighteenpence a time. Mr. William Bennett has now charge of this really practical benevolent institution. Strangers pay more than the inhabitants to have their nets tanned, and fishermen from Rye and other parts of the neighborhood avail themselves of Sir Eliab Harvey's foresight. This great man was a native of Folkestone, and there is now an "Harveian Institution" in the town. His mother was buried in the parish church of St. Mary and St. Eanswith. A tablet to her memory has lately been removed during the alterations; it luckily fell under the notice of one of our most learned London physicians, who has caused it to be carefully preserved in an appropriate place in the church.

The migration of animals may be instanced as one of the appointed laws whereby man derives benefit, directly or indirectly, from the obedience of the created beast, bird, or fish, to the will of the Creator. Unconsciously they follow the promptings of their instinct, and, in so doing, they not only derive pleasure to themselves, but also play the special part allotted to them in the chain of animal life, the links of which are so united that the acts of individuals finally become conducive to the well-being of the whole.

The miraculous migration of quails in the wilderness of Sin was, we read, brought about by the direct interposition of Providence for a special purpose, viz. the support of the starving tribes of Israel, already murmuring at the loss of the fleshpots of Egypt.

Who can dare say that this same influence over the animal kingdom does not continue at this very day, though, from its regular occurrence, we are familiarized with what is in reality a continuous miracle? Who can sufficiently admire the beneficence which ordains that countless myriads of herrings and other fish should annually visit our shores—the sea yielding a crop of fish not only good for human food, but even exceedingly palatable to the taste? What is this but the fulfilment of a law which ordains that things animate and inanimate shall all pay their tithe to man, and help support him in his position as the head of created beings?

There are two theories as to where the herrings come from. The old story—promulgated, if not originated, by Pennant—was, that these mighty armies of fish came from the far North, from those inhospitable, ice-locked regions, where it will be impossible for our race ever to attempt to thrive. Pennant says:—"The herrings begin to appear off the Shetland Islands in April or May. This is the first check the army meets with in its march southward. There it is divided into two parts; one takes to the east, the other to the western shores of Great Britain; others proceed towards Yarmouth, the great and ancient mart of herrings; they then pass through the British Channel, and after that in a manner disappear," &c. &c. The western division he describes as being divided by Ireland into two brigades, one of which runs down each side of the coast. The more modern idea is, that the herrings come up out of deep water, in order to deposit their spawn in the shallow water, where it may become vivified by the light and heat, which would not reach it in the deep sea. They remain some time near the shore before they spawn, in order, as it is supposed, that the ova in their bodies may become mature; and this is the time when they are best fitted for food.

The proprietor of the Folkestone Herring Hang strengthened this theory when he told me that "at the beginning of the season all the fish have roes—towards the end they are all 'shotten,' *i. e.* they have no roes." The shotten fish are worth much less than when they have roes. We ourselves must have remarked that the herrings we buy in the London market, when they first come into season, are quite bursting with a mass of eggs. My Folkestone friend was, however, of opinion that herrings "were a fish of passage" He says :—"They catch them in June at Shetland, about October off the North Foreland, in November at Folkestone, at Hastings soon after that, and then more to the southward, off the coast of France, up to the month of Febuary." I give this for what it is worth. It is possible that this migration stream towards shore begins at the north part of England, and gradually runs along the coast to the southward; and it is a very pardonable mistake to suppose that the line of march of the fish is down the Channel, and not from the centre of the Channel towards the sides.

My friend, Mr. Roberts, having read the above, wrote to me as follows :—

"Herrings, sprats, mackerel, and pilchards, have a deep-sea habitat; this they leave under the operation of a certain law, and come to the oxyginated surface water.

"They do not migrate from the distant North,

as Tennant asserts. It is true that they appear at different localities at various times. Let us take the herring. It strikes in to the north of Scotland in early autumn; it is early winter before the shoals of this fish appear at Dover and south coast, 600 miles from the great fishery of Wick. This looks, at first view, very much like migration from the North, but it is not so. Whenever the fish strike in first (I say from the deep, their particular habitat), they are full of roe, or, as it is expressed, 'in season.' This does not continue long, and the fishermen yawl or mesh in their nets only 'shot herrings,' lean and low priced in comparison to what they were a week or two before. This is repeated at every locality along the east coast of England and Scotland.

"Were the herrings of Wick to come to us in the south, they would spawn by the way; and what would the Yarmouth bloaters be? A Worthing boat sets out to the north for herrings. The skipper thinks not of meeting the herring army, but he goes eastward, and depends upon accounts of what is doing off many ports. He looks for his penny letter, sent by some friend, advising him whither he shall steer—northward or southward—and where the gulls are, and where the herrings full of roe abound. There is no such thing as a continuous stream of fish. They come in detached shoals. The necessity for early information, so that the fish do not stay in the fiords uncaught and shedding their roe, has caused the Norwegians to establish an electric telegraph, so that the herring fleet may sail at the earliest appearing of the fish to the fiord or fiords where the full-roed herrings are.

"The sprat is found all through the summer in the stomachs of our larger fish. I cannot learn that it is so with the herring, or the mackerel. The mackerel comes in from the ocean; so does the herring. The salmon's habitat is known to the seal, who leaves the marks of his teeth on the backs of the fish.

"As to the law which makes the deep-sea fish stir, this we know, that they are obedient to it. Mackerel strike in upon the shores of the west in great numbers at times in the autumn Temperature may play an important part.

"Lake fish become at one season restless, and migrate to the other end of the lake. The subject is a grand one."

When the herrings come off Folkestone, the boats all go out with their fleets of nets, "yawling," *i. e.* the nets are placed in the water and allowed to drive along with the tide, the men occasionally taking an anxious look at them, as it is a lottery whether they come across the fish or not. It is the received opinion that the fish rest at the bottom of the water. One of the men stated that he had seen them resting themselves once, and once only; when looking down into the water which happened to be very clear, he saw "thousands of fish, all quite stationary and resting, their noses being on the sand at the bottom, their bodies off the ground." The nets sometimes drift a long time, catching nothing, and all of a sudden "the fish rise like a flock of sheep," and the nets are full. The fish have, moreover, stated times for rising, and these are, "at the rising or setting of the moon, and at daybreak, just before sunrise."

All anglers well know how sensitive fish are to atmospheric changes; the above facts, noticed by men whose living depends upon their powers of observation of the habits of fish, are an additional proof (were it required) of the influence of meteorological causes upon the movements of the inhabitants of the waters, both salt and fresh.

Strange things are seen and heard by the fishermen, when drifting silently along, far out at sea, through the murky darkness of midnight. My friend, Mr. Roberts, of Dover, informs me that a gentleman was lately out on one of these occasions, on a remarkably dark night, in order to see the process of catching the fish. All of a sudden, a curious rushing and a rustling sound was heard over and above the boat, accompanied by low musical twitterings; this phenomenon was not continuous, it passed away, and in a few minutes was again repeated; meanwhile all was darkness and mystery. He asked the man what strange noise that was; they shook their heads in ignorance; they did not know what it was, whether earthly or unearthly; they evidently regarded it with somewhat more than awe, and were unwilling to converse about it; their dread of it was great; all they knew was that it was the "Herring Piece." I asked one of the oldest fishermen in Folkestone about this. He had heard the noise often, he told me, but that the Folkestone name for it was the "Herring Spear." "I likes to hear it," he said; "we always catches more fish when it is about." And what is this "herring piece," or "herring spear," after all? The courage of the sturdy fisherman, who will face the gigantic waves of a storm from the south-west, and bravely run his boat down into the boiling surf to render assistance to the Indiaman that is firing minute guns on the Goodwin Sands, is appalled and trembles at the rushing noise of the wings and the musical call-notes of the flocks of those pretty little birds, the redwings, who choose a dark still night to wing their flight across the Channel to the British shores. Their time of migration is always about the herring-fishing time, and hence the noise caused by them in their flight is always associated with these fish. No harm follows the "herring piece" to the listener; not so with the "Seven Whistlers;" woe betide those who hear this evil-portending sound; "I never thinks any good of them," said old Smith; "there's always an accident when they comes. I heard 'em once one dark night last winter. They come over our heads all of a sudden, singing 'ewe-ewe,' and the men in the boat wanted to go back. It came on to rain and blow soon afterwards, and was an awful night, sir; and, sure enough, before morning, a boat was upset, and seven poor fellows drowned. I knows what makes the noise, sir; it's them long-bill'd curlews; but I never likes to hear them."

Besides hearing these strange sounds, the poor fisherman often sees the "Composant." As he sails along, a ball of fire appears dancing about the top of his mast; it is of a bluish, unearthly color, and quivers about like a candle going out; sometimes it shifts from the mast-head to some other portion of the boat, where there is a bit of pointed iron, and sometimes there are two or three of them on different parts of the boat. "It never does anybody any harm, and it always comes when squally weather is about." I know not the origin of this curious term "composant;" the nearest I can get to it are the French words, "*coin blasant*," burning or shining corner; or, better still, from the Italian "*corpo-santo*." Englishmen are not good hands at inventing names; and I think the Folkestone people most likely picked up the word from the Frenchmen whom they met out at sea in pursuit of herrings as well as themselves.

Besides this "composant," the "water-burn" is very frequently met with; and by this is meant the very peculiar luminous appearance of the sea, when the waves seem to be formed of liquid phosphorus, and to leave a coating of phosphorus upon all they touch. "Water-burn" is much disliked by the herring yawlers, as the cunning fish can see the net and will not go into it. When comfortably seated at dinner in a warm room, with a dish of these fish before us, let us think of these poor fishermen, who risk their lives these long winter nights, watching their drift-nets in solitude, cold, and silence.

There is at Folkestone harbor a long flight of stone steps, which in the warm summer days becomes a sort of fishermen's club, where, when work is over, they meet and discuss the affairs of the harbor. Close by these stairs I espied, half buried in the ground, a gigantic bone, which these men told me was a whale's "rump-bone." Upon examination it turned out to be, not what they described, but the back part of the head of a large whale. It was dug out of the "lug sand" (the sand where the lug-worm is found), when they were building the pier. Though much knocked about, I could distinguish in this bone the brain cavity and the holes for the exit of the nerves, some of which must have been as large as a man's thumb. This was not the head of a fossil whale, but of some whale that had probably died out at sea, been washed ashore, and buried in the mud. None of the fishermen knew anything of its history; it might have been one hundreed or five hundred years old, but it was not a fossil in the true acceptation of the term.

It is not such a very long time ago that whales (the true or "right" whale) were common in the English Channel: in the time of Edward the Third, A.D. 1312-27, they were frequently seen and captured; and every now and then, even in the present day, they appear off our coasts.

In olden times a curious law existed, that when a whale was taken on the British coast it should be divided between the king and the queen, the head only being the king's property, and the tail the queen's; the reason for this distinction being, as assigned by our ancient records, "to furnish the queen's wardrobe with whalebone."

I never saw a whale alive; but they have been seen alive near Folkestone, and I heard several stories about them from the fishermen.

Some twelve years ago a whale appeared off Hythe. The inhabitants turned out and attacked him with guns, scythes, and all sorts of weapons; they seriously wounded him, but did not kill him. He was afterwards found floating dead out at sea, and towed into Folkestone harbor. They sunk a barge under him, and, as the tide went down, allowed him to sink into the barge. The barge, whale and all, with his tail hanging over, was afterwards towed up to London by a steamer, and was sold for 40*l.* for oil.

Once upon a time, as Mr. Smith, one of the most experienced of the Folkestone fishermen, tells me, a large whale appeared off Weymouth, and was seen by several fishermen. The affair was talked over at night in the public-houses, and one of the company, who happened to have a new seine-net which he had never used, was much taunted about it, and he was dared to go and net *this* whale with his new net. At first he took it as a joke, but, under the influence of beer and the chaff of his comrades, he stamped his hand on the table and said, "Well, d— me, if I don't go and shoot the net after him, catch or no catch." Accordingly a sentry was posted, and the next morning the whale was signalled as being in the offing. So the owner of the new seine put it into the boat, and, rowing quietly along, shot the net round the unsuspicious whale. At last Mr. Whale put his nose into the net, and feeling something strange, charged against it, dragging men, boats, and all along with him. He then plunged and dived, and, ultimately taking the new seine-net, rolled about his body, right away with him, in spite of all the fishermen could do. They looked after the whale, who had gone off with the net, much as an angler looks into the water when a fine fish has escaped from his hook; but, however, the whale was gone, and the would-be captors rowed home disconsolate and whaleless.

Some three or four days afterwards, as a coastguard was going his rounds in the dead of the night, he saw a huge black mass come rolling in with the tide; it looked like a wreck, yet it was not a wreck, for a wreck has not a tail wherewith to flop the water as the object had. The coastguardsman waited till the tide turned, and as it went down he got near to this strange object, which had got hard and fast among the rocks. He then saw that it was a whale, and, what was exceeding strange, the whale had a net entangled round about him in the most complicated manner. "First come, first served," said the coastguardsman to himself, as he pulled out his knife and cut two great slashes in the whale's fat sides, during which operation (mark it, O reader) the *whale kicked and evinced signs of life.* The finder then shut up his knife and posted off with the news. Of course, as there was a net round the whale, his identity was established directly, and the owner of the net claimed the carcass because his net had caught him; the coastguardsman claimed it because he had found him. Meanwhile, when the dispute was still going on, the lord of the manor put in his claim, as it was found between high and low water-mark, gained it, took possession of the whale, cut him up and boiled all the oil out of him, getting forty barrels, worth a lot of money; and there the matter ended.

Some weeks afterwards, as the coastguard was sit ing on his "donkey" (the term applied to the portable stool used by these men,) a respectable-looking gentleman walked up to him, and said, "My man, don't you recollect the whale that you found hereabouts some time since?" "Yes, sir," said the man, "it was me as found him." "Well, now, can't you recollect whether, when you cut him (as they tell me you did,) he kicked and winced under the knife?" "In course he did!" was the answer; "he nearly knocked the knife out of my hands with his tail." "Well, then," said the old gentleman, bristling up all of a sudden, "now, I am lawyer, and mind that you tell the same story to-morrow, sir; for, as sure as to-morrow comes, you will have to *swear* in court." On the morrow the coastguardsman swore that the whale was alive when first he saw him on the shore, and that he knew it by the knife test, as stated above. It was now the lord of the manor's turn to sing small, for he could not claim a thing if cast up *alive.* He had to refund the money he got for the oil, having taken all his trouble for nothing; so that, after all, the owner of the new seine caught his whale, got his new net back, and nearly a hundred pounds besides.

I mentioned this fact to Mr. Marder, of Lyme, a naturalist and a great sportsman. He told me a curious circumstance of whales' bones. One day, when out shooting, he found in the middle of a ploughed field a great round thin bit of bone, which, as an anatomist, he at once knew to be from off a whale's vertebræ; but how did it get into the ploughed field? He afterwards ascertained that the farmer to whom the field belonged had bought from a fisherman the skeleton of a whale, and had pounded it up for manure. The fisherman had shown the skeleton for some time, but soon got tired of the oily bones and the exhibition, and was glad to sell it to the farmer for manure; and this accounted for the whale in the ploughed field.

Whales' bones get to odd places. I know a garden at Abingdon, in Berkshire, where the entrance to a pretty artificial grotto is under a porch of ivy. The ivy, I found, is supported by the lower jaw-bones of a gigantic whale. How they got it there I could not ascertain. In a garden at Clapham I once saw a chair that had a very anatomical look about it; no carver would ever have thought of that shape for a chair, thought I. I examined it, and found that it was one of the vertebræ or back bones of a whale, painted green; it made a capital seat.

In the neighborhood of Hull, and also Newcastle-on-Tyne, it is, according to Birdcatcher, "very common to see the huge jaw-bones of whales used as posts for field or even for approach gates, and very good and durable ones they make, whether painted or not."

The rib of the Dun Cow at Warwick, and the gigantic rib at St. Mary Redcliffe's Church at Bristol, are the bones of whales. There is suspended in the yard of the Royal College of Surgeons a gigantic blade-bone, with an anchor painted in gold upon it, and Calvert and Co's Entire underneath. This bone, Professor Quekett informs me, formed, for half a century, the sign of a public-house at Portsmouth, where they sold Calvert's beer. The blade-bones of whales are not uncommonly seen at the present day in the bone-shops of London. There is the top of a whale's humerus or arm bone hanging outside a bone-shop in Hammersmith, with the following inscription:—"The largest bone ever known: weighs 128 pounds." I have seen another in a shop near the Vauxhall Road; also a large scapula in the "Seven Dials;" they seem placed, not for sale, but to attract attention. The immense importance of whales' bones to the poor Esquimaux is too well known to need comment.

The whales seen about the British seas are generally stragglers from the northern waters; but it sometimes happens that a whale from the southern regions pays us a visit; however, this is a very rare event. As my readers are probably aware, the great Sperm or Spermaceti whale is found south of the equator only, and, according to Maury's "Physical Geography of the Sea," there is a line (marked in his map) across which the sperm whale cannot pass. Nevertheless, I have a well authenticated case of a sperm whale appearing in the Bristol Channel. I have also a good engraving of a whale which was cast ashore near Antwerp, A.D. 1576. This creature is also a large sperm whale. There is a figure of a man measuring it, and the result of his observations are (as it were) given in the inscription on the plate: "Ein grosser Wallfisch von 60 schuch lang und 41 schuch hoch." The costumes of the people about are most quaint and curious. There was evidently great excitement about it, as streams of people are seen in the distance, and an old-fashioned coach, with the horses galloping. There is a booth erected, in which a man is playing the fiddle, and people drinking. There is no mistaking a sperm whale when once seen; the form is different from the true or "right" whale; and he has teeth in his lower jaw, whereas the right whale has "baleen," improperly called whalebone. These sperm whales' teeth are very common in curiosity shops and in private collections, and they often have figures of sailors, ships, anchors, &c. carved upon them. Many of the so-called "ivory" handles of ladies' parasols are made of rods of bone cut out of the solid "teeth-armed" under jaw of this monster of the deep. The "engine-turned appearance," as seen in all sections of ivory, such as an opera ticket, upon this so-called ivory handle will be absent; and if so, the owner of the parasol may console herself that she is holding in her hand, not a portion of an elephant's tusk, but of a whale's jaw. Curious it is that in this respect the most gigantic of land animals and most monstrous of marine creatures should become subservient to the hand of beauty.

Talking of whales to a friend well known as one of the first of our English naturalists and ornithologists, I learnt from him the following story, which he has kindly allowed me to place on record:—

"Nearly thirty years ago, as a French vessel was going down the Thames, her captain saw a huge object floundering about on a sandbank near the Nore. A boat was put out, and this object was ascertained to be a large specimen of a sperm whale which had got on the sandbank, and was gradually being left high and dry as the tide went down. They waited till the water allowed them to get to the whale, who appeared *almost* dead, and then, passing a rope from the ship, they tied it tightly round him above his tail, and returned to the ship. As the tide rose, the whale floated, the ship set sail, towing along the whale in high glee, thinking in their simplicity that they were going to take him as a prize into Calais harbor. Mistaken Frenchmen! They had not got far before the whale showed symptoms of returning life, by wriggling, twisting, and not following the ship in the orderly manner becoming a *dead* whale. At length the dragging through the water and the pulling at his tail roused him up completely; he came *quite* to life, and threatened to damage the ship. The Frenchmen, finding they had caught a Tartar, cut the rope as it passed over the stern of the ship, and away went the whale one way, back again towards the mouth of the Thames, with a great, long, thick rope still tied to his tail and dragging behind him; the ship went the other way, towards Calais, having lost their prize.

"The whale was seen occasionally for some days after this, still with the rope on his tail. At last he foolishly came into Whitstable Bay. The men in the oyster-boats saw him, and, by shouting and splashing, forced him up into shallow water, where the tide left him. They then wisely made sure of him by killing him outright. No pig-sticking knives were big enough to cut his throat, so they fixed scythes on to poles, and pierced him till he died. Then came the question, what was to be done with him? and at last a deputation of fishermen went up to Messrs. —— and Co. in the City, the great dealers in whale-oil, whalebone, &c., who gave the fishermen 80*l.* for their capture, and sent down men, blubber-knives, iron pots, &c., to boil him up. At the same time an offer was made to a London society, to which my friend was attached, to send down and take what parts of the whale were wanted for anatómical and scientific purposes. Of course he went down immediately to Whitstable Bay, and there found this gigantic sperm whale, between seventy and eighty feet long, surrounded by workmen digging and cutting the blubber off his huge carcase, and filling the surrounding atmosphere with a most overpowering effluvium which was anything but pleasant. Everything was on such a gigantic scale, that the dissection could not be carried on without the aid of horses; so when the blubber was all taken off the upper side of the whale, my friend harnessed the horses to the ribs, and hauled them out one by one, thus exposing all the contents of the body. He then carefully descended into the gigantic mass of anatomical horrors, and took out what parts he wanted. This service was not, however, done without danger, for when dissecting the enormous heart his foot slipped and he fell into one of the cavities of the heart, his feet passing down into the great artery, the aorta. Assistance was luckily at hand, or he would have slipped right down into this huge pipe, and might have met with a fatal accident. To show the narrow escape he had, he subsequently cut rings out of this aorta, and found he could pass them, without stretching, over his head and shoulders right down to his feet.

"Nothing daunted by the smell, or other disagreeable circumstances, he continued his dissection till he had made a valuable and interesting collection of preparations. One morning, when he was still working away, and the men boiling or 'trying down' the oil, a very official-looking personage came up, and in a haughty manner said: 'Leave off work, all of you, directly; do nothing more to the whale. I claim this *fish* (he was no naturalist, for a whale is a warm-blooded animal, and no fish) in the name of the *lord of the manor*, and nothing whatever must be touched or removed.' My friend had just taken out the whale's eye, which exhibits remarkable structures, and, offering it to the new comer, said: 'Well, I suppose the lord of the manor does not want the whale's eye?' 'You may keep *that*, sir,' was the answer. This disagreeable intruder then departed.

"While the dissectors, both anatomical and commercial, were recovering their wonder, from another quarter of the compass, lo! another Jack-in-office, who stalked into the assembled crowd, and with a wave of the stick proclaimed: 'Touch nothing—move nothing—I claim all this in the name of the Lord Warden of the Cinque Ports.'

"Vultures smell the stinking carcase from afar; so did it seem that the smell of this unfortunate whale had found its way so far as London, and even to the Crown authorities, for yet another human vulture arrived on the wings of the wind, and, spreading his official papers over the putrid mass pronounced, 'This is a royal fish, and I claim it in the name of the Crown.' Here, then, was a pretty mess, both literal and physical; the three claimants set to work disputing with each other for possession; the fishermen quietly wiped their oily hands, packed up their traps, and adjourned to the public houses to drink out their eighty pounds and talk over the matter. My friend, seeing nothing could be done, packed up his whale's eye, and went back to London in an oyster-boat, as he carried such aroma about with him that they would not take him in the coach. The three vultures then departed to their respective nests, and transferred the cause to the lawyers, who managed to spin out their disputes for a whole year. In the meantime a fourth claimant, and that a most powerful one, appeared. Father Neptune, seeing that mortals would not take advantage of his gifts, ordered his white-capped servants, the rolling sea breakers, to remove his present; they did so, bit by bit, rolling back into the bosom of the ocean, blubber, oil, bones, and anatomical preparations, except the huge jaws, which were saved, and, if I mistake not, are now in a garden at Canterbury. So that by the time that the lawyers had finished their learned arguments, there was nothing whatever left of the sperm whale, about which they had been so foolishly disputing."

The skulls and bones of whales have often given rise to many wonderful stories; but nothing can beat the story of a marvellous skull that was found in Australia, and of which an account from the *Sydney Morning Herald* was, in 1847, sent to my father for his opinion. It runs as follows:—

"THE APOCRYPHAL ANIMAL OF THE INTERIOR OF NEW SOUTH WALES.

"*To the Editors of the Sydney Morning Herald.*

"GENTLEMEN,—I need scarcely observe to you that this colony is distinguished by the most grotesque variations of the customary phenomena of nature; birds without wings scour our plains, and marsupial quadrupeds, with claws on their fore paws and talons on their hind legs, like birds, hop on their tails; the moles lay eggs, and have duck's bills; we have birds with brooms in their mouths in place of tongues; fish for which it is utterly impossible to find a place in the existing systems of scientific men; and salt growing in perfection on the bushes of our forests. Until lately it was supposed that nearly all our quadrupeds belonged, or were intimately related, to the *glires* of Linnæus; but whilst it was generally known that at least two-thirds of the Australian quadrupeds made their way by springing in the air, it has been but lately that rumors have reached us of a huge animal of the *feræ* order disporting in clumsy gambols, and inhabiting the waters of the lakes and rivers of the interior. These rumors have, however, begun to assume a more certain form; and, inasmuch as during my recent trip on the banks of the Lachlan and Murrumbidgee, and through the Murray district, many details in reference to this apocryphal animal were given to me, I will, with your permission, lay before your readers such particulars as I have been enabled to collect.

"The Murrumbidgee blacks assert that a large animal, '*big as him bullock*,' exists in the lakes of that district; they describe it as having a head and long neck like an emu, with a thick mane of hair from the top of the head to the shoulders; four-legged, with three toes on each foot, which is webbed: and having a tail like a horse. They call it the *Katenpai*, whilst by the Watta Watta tribe (who similarly describe it) it is called *Kyenpate;* by the Yabala Yabala tribe on the Edward River, it is known as the *Tunatbah;* whilst the Burrula Burrula tribe call it *Dongus.* I have been informed that the blacks on the Great Carangamite Lake, in the Portland district, describe a similar animal which they call the *Bunyip;* and I have heard various accounts from white men (shepherds and others) who profess to have seen the animal at its gambols in the water. But the following incident has been productive of so tangible a result that, however I may have doubted the exaggerated narratives of some of my informants, I cannot but conclude that some large animal, with which we are yet unacquainted, really exists in the districts I have named.

"Mr. Fletcher, who resided on the Lower Murrumbidgee, was told by a tribe of blacks that they had some time previously killed a *Katenpai*, on the banks of a lake near the Murrumbidgee. It must be observed that the blacks have a great dread of the animal, and avoid bathing or fishing in the waters where they assert that it exists. They assured Mr Fletcher that the remains of the creature would be found in the spot where they killed it; and, although doubtful of the fact, that gentleman proceeded to the place minutely described by the blacks, and there found a large portion of the skull of some animal, which, to all appearance, had not been dead for any great length of time. No traces of any more bones or other remains could be discovered, but enough was found to prove the existence of the supposed fabulous *Katenpai*. Every black to whom the skull was afterwards shown agreed that it belonged to the dreaded monster of the lakes; and in order to give your readers as accurate a notion as is in my power of all that can be gathered from Mr. Fletcher's discovery, I will request their attention to the following rude sketches of the skull, which was afterwards taken by Mr. Fletcher to Melbourne, and where it will doubtlessly receive the most careful examination from skilful comparative anatomists."

[Here follows a description and drawing]

This slip was sent to the late lamented Mr. Broderip for his opinion, which is as follows:—

"I return your Sydney slip with thanks. If the artist who drew the figures be anything of a draughtsman, the skull looks marvellously like that of a calf. The *Katenpai*, appears to be a near relation of the Water Bull of some of the Scotch Lakes. "Ever yours,

"W. J. BRODERIP."

Professor Owen also wrote his opinion to my father, and he has kindly allowed me to insert his letter in this place:—

"The opinion which you have received from Broderip I believe to be near the truth. It was founded, however, not on the cuts in the slip from the newspaper, which are problematical enough, but on a pencil sketch the size of the skull itself, brought to me a few days ago by a gentleman from Port Philip, who has seen the specimen. That sketch showed the crowns of the teeth of a young ruminant just peeping out of the sockets, and left little doubt in my mind that the story had no other foundation than the skull of a calf.

"Ever yours,

"RICHARD OWEN."

This "tale of a Bunyip" shows, what I have often experienced myself, how readily uneducated and illiterate workmen and laborers (and savages in particular, in the case of travellers,) will instinctively endeavor to impose impossible stories upon the "gentleman as looks after curiosities." They generally, however, defeat their own object by stretching the truth to such an extent that they are obliged to laugh at their own stories; and it is most amusing to strip off one by one the bandages of lies, in the centre of which a very small bit of truth indeed lies buried. Good humor and beer should be the instruments employed in this dissection.

There is a natural curiosity that occasionally

turns up in London in the shape of a mermaid, and I have been lucky enough to have seen one. The greatest of British naturalists (Professor Owen) having, in August, 1858, kindly informed me that there was a specimen of a "merman" on exhibition in London, and that he wished me to examine and report on it, I hastened (honored with the request) forthwith to view this monster of the deep.

In the back parlor of the White Hart, Vine-court, Spitalfields, high and dry upon a deal board, lay this wonderful object—hideous enough to excite the wonder of the credulous, and curious enough to afford a treat to the naturalist. Such a thing as a merman or mermaid of course never really existed; I was therefore most anxious to examine its composition, which, by the kindness of the landlady (a remarkably civil woman), who removed the glass which covered her treasure, I was enabled to do. The creature (a gentleman, not a lady specimen of the tribe) was from three to four feet long. The upper part of its body was composed of the head, arms and trunk of a monkey, and the lower part of a fish, which appeared to me to be a common hake; and the head was really a wonderful composition: the parchment-like, hideous ears stood well forward, the skin of the nose when soft had been moulded into a decided specimen of "the snub," the forehead was wrinkled into a frown, and the mouth "grinned a ghastly grin;" the curled lips partly concealed a row of teeth which in the upper jaw were of a conical form and sharp-pointed, taken probably from the head of the hake, whose body formed the lower part of our specimen. The lower jaw contained these fish's teeth, but conspicuously in front was inserted a human incisor or front tooth, and a vacant cavity showed that there once had been a pair of them. These were probably placed there to show the "real human nature" of the monster. The head had once been covered with hair; but visitors, anxious to obtain a lock of a merman's hair, had so plucked his unfortunate wig that only a few scattered hairs remained; the relic-seekers are now, therefore, ignorantly treasuring in their cabinets hairs from the pate of an old red monkey.

The eyes, sunk deep in the sockets, are formed of round bits of leather, with the pupils marked in black paint; and altogether the features of the merman are those of a disagreeable old man, who was trying *not* to laugh. There is no portrait of the merman tribe in "Bell's Anatomy of Expression," and a portrait of our Spitalfields friend ought really to find a place in the next edition.

The arms, long, shrivelled and gaunt, were placed in an easy position, as though the owner was kissing its hand to the spectator, and the soft parts having receded from the nails left them long and projecting, like a bird's claws. The chest of the monkey had hardly been big enough to hold the shoulders of the fish, so it is extended with a cage of wire, which also gives the appearance of ribs. The waist is very much larger than the chest proper; from which fact we may learn that the fashion of tight-lacing was *not* derived from the mermaid family.

The fish (neatly stuffed) was placed with its

belly outermost, so that its back fin formed a continuation with the back of the monkey. The junction was cleverly managed, and the tail part was gracefully curved to the left, like the heraldic pictures we sometimes see on coats of arms, &c. The merman was placed on his back; but his proper position is evidently erect, for if he stood up on his tail he would have a much more imposing appearance. The history of it is, that it was bought at a sale of old furniture, &c., of a certain old Mr. Ellis, of whom all I could learn was, that "he bought and sold for the East India Company;" but whether he bought and sold tea, silk or mermaids, I could not ascertain. The present proprietors think it was sent to Mr. Ellis from "foreign parts," on account of the foreign look and spice-like smell of the material which formed its bed in the box with which it was bought. Rumors, however, came to my ears that it is a "London-made" merman, and that a certain bird-stuffer in the west end of town, several years ago, made a pair, a merman and a mermaid, and sold them for exhibition. Immediately upon hearing this an idea struck me that I knew the whereabouts of Mrs. Mermaid. She also is in London, and but a short distance from her husband.

In an old curiosity-shop, in the west arcade of Hungerford Market (where they sell poultry), I found my lady, looking as pretty as ever, under her glass case; and the proprietor kindly allowed me to inspect her. About half the size of her partner—she too is formed of a monkey and a fish—but not put together quite so artistically: her head is too bullet-shaped, her eyes decidedly glass doll's eyes, her teeth no teeth at all, but a small bit of bone just cut into notches, and which interfere much with her personal appearance and detract from her interesting look; but to make up for this, her hair is longer, and her chest, &c. exceedingly well developed. She is fastened upright by means of the curved portion of her tail, and smiles gracefully through her dusty glass house. Her history, as told me by the proprietor, is curious: she came from Yankee land, and was exhibited years ago at the Egyptian Hall, forming one of the first, if not the very first, exhibitions in that place. She was sold to two Italian brothers for 40,000 dollars, and there was a Chancery suit about her, as one of the partners wished to prevent his brother exhibiting her. Her age is certainly forty-five years, as the present owner could trace her during this period—how much older she may be it is rude to inquire, considering her sex.

Mermaids were, I believe, not very uncommon exhibitions in days gone by; and they may be still seen occasionally at country fairs, &c. The good folks of England are getting every year more and more educated, and mermaids do not take so well now as formerly, when pack-horses performed the part of railways, and horn-books composed the village library. Still the idea of a mermaid can be traced to very remote ages. The ancient Greeks had a presiding nymph over their favorite fountains and streams. The bold sailors of Æneas kept well to leeward of "the rocks of the sirens, white with scattered bones;" well they knew the stories of the three sister sirens, "Parthenope, Locosia, and Ligea," who, presenting the form of a beautiful woman above, a fish below, with soft music enticed the mariners to shipwreck and destruction.

For a belief so universally prevalent there must be a foundation of some kind. Sailors, always superstitious, and employed for the most part in an occupation of solitude and apprehension, would easily convert into a human form the sudden appearance of a seal, walrus, &c. coming up from the depths of the ocean to breathe and look about. Thus Dr. Scoresby, the celebrated Arctic traveller, writes:—"I have myself seen a sea-horse (a walrus) under circumstances that it required little stretch of the imagination to mistake it for a human being; and the surgeon actually reported to me that he had seen a man with his head above the water." Let the reader look at the big walrus at the British Museum, and if he has a large acquaintance and a fertile imagination, he will very possibly recollect some one of whose face he is reminded by the features of that animal. Sir W. Jardine's opinion, is, "that it is generally some species of seal, very frequently the barbata or haaf seal, which, from its more solitary habits, has given rise to these mermaid legends." No one who has watched the seal at the Zoological Gardens can help being struck with its very human face as it gazes out of the water; and even when it turns round and dives, the lower part of its body gives the appearance of a man diving with his legs tied together. But the most human-like of sea creatures are the Dugong and Manatee, which are found in the warmer parts of America and its islands, and also in Western Africa and Australia; their skulls (there are several in the Royal College of Surgeons, Lincoln's-inn-fields) resemble that of a man with a very long nose; their mammæ are placed in the same position as in the human species; and they have very free use of their anterior extremities, which they use for progressing, nursing their young, &c.

The Portuguese and Spaniards call them by a

name which signifies "woman fish," and the Dutch call them "Baard mannetje," or little-bearded man, as they have a quantity of thickset hair round their muzzles. These creatures are herbivorous; after cropping the herbage at the bottom of the water, they rise up suddenly, thus exposing the upper part of their bodies out of the water; at the least noise down they go again; having firmly impressed poor Jack on the look-out from the foretop that he has seen a mermaid, Jack comes home with his story, and takes care to tell it; and if the ingenious Taxidermist can't catch a merman or mermaid, there is no reason why he should not make one. The "lines" are laid down by Jack—"half a man half a fish," are the orders; a monkey's body makes a capital figure-head, a fish's body a capital stern; a glass-case keeps off inquisitive fingers, and the merman and mermaid are ready for the "walk-up-ladies-and-gentlemen" orator, who not unfrequently reaps a good harvest from those who go to see because they do believe, and those who go to see because they don't believe.

I lately heard of a speculator giving a very large sum of money at the Cape of Good Hope for a mermaid. He exhibited it; and when he had got all his money back, and something over for his trouble, his curiosity could no longer be restrained. He cut up the mermaid publicly, and discovered its composite nature.

I attended Mr. Barnum's lecture on "Humbug," and the following are my notes of what he said of his celebrated Fejee Mermaid. He defined "Humbug" as "The art of attracting attention, whether the article is good or bad;" and his mermaid story exemplifies his theory. He bought the mermaid, which was being exhibited in Watson's Coffee-house, London, for a shilling, and then he (Barnum) showed it in his museum *for nothing*, and yet made money. He had an elaborate and really beautiful picture painted, which he hung outside the museum; the picture represented three lovely creatures with beautiful long hair, the traditional looking-glass and comb, &c.

disporting themselves in a fairy-like submarine grotto; but he did *not say* his mermaid was like those in the picture. Attracted by the picture and notice, " A mermaid is added to the museum, —*No extra charge*," thousands paid to go in, and then they saw a " hideous, shrivelled-up old mummy; and if people were not satisfied with the mermaid, they had their shilling's worth in looking at the rest of the museum." Mr. Barnum confessed that he did not pursue his studies in Natural History *too far, or he might learn too much.*

Much of the small-shot of wit and pun has lately been fired at an extensive Pavilion erected upon the area of the King's Mews at Charing Cross (probably with Carlton House in their recollection). Epigrammatists have called it the " Palace of the Prince of W(h)ales," others, " A tub for a Whale;" but we must assure them that the stupendous occupant of this handsome building is entitled to their courteous attention. At Ostend this occupant of the deep created what journalists call a sensation, and our gay neighbors made its capture the occasion of three days' "fetes," with a host of allegorical and processional accompaniments, which are detailed in a "Mémoire pour servir" of the whole affair. The proprietors next visited Paris, and there pitched their Pavilion in the Place Louis XV. Thence they journeyed to London, where we hope the exhibition will receive all the encouragement it merits, &c.

The skeleton is placed in the area or pit, and within the ribs is a stage, to which the visitor ascends by a flight of steps. Here are tables and seats; on the former we found three volumes of Lacepede's Natural History, and a folio album, with epigrams, poems and other small wit of visitors at Ostend and Paris. One specimen, *à conundrum*, signed by two illustrious gladiators, is as follows: "Why should we be mourned for, if killed by the falling of the bones of the whale? We should be be-wailed." "Why is the Whale like Noah's Ark? Because it contains an Ass."

Altogether, "The Pavilion of the Gigantic

Whale is one of the pleasantest places we have visited this season."

Again, in May, 1836, were exhibited, at the Cosmorama, Regent-street, London, "The wonderful remains of an enormous head, 18 feet in length, 7 feet in breath, and weighing 1,700 pounds. The complete eighty-eight bones were discovered in excavating a passage for the purpose of a railway, at the depth of 75 feet from the surface of the ground, in Louisiana, and at the distance of 160 miles from the sea." My father has annoted on this handbill, "Balcena Mysticetus, recent and oily."

In the month of March, 1857, there appeared in the *Times* an advertisement for a vacant bit of ground whereon a whale might be exhibited. I watched anxiously for the result, and with success; for shortly I read another notice to the effect that the whale had arrived, and was now in the Mile-End Road, Whitechapel, near the King Henry the Eighth public-house. That same day found me on the top of a Bow and Stratford omnibus, the conductor promising to set me down "at the whale."

The admission fee of sixpence being paid, I entered a tent, and for the first time in my life enjoyed a full and uninterrupted view of the monster. I had expected to have seen a skeleton; but, instead, the proprietor has preserved, stretched on a framework, the skin entire. The head remains attached with the bones, whalebone and all complete; so that it was a stuffed whale I went to see, and not a skeleton—none the less interesting for that. It rarely happens that Londoners have a chance of seeing a specimen of the largest animal in creation. Pictures certainly convey an idea in a whale; but to have a notiou of its huge bulk, the thing itself must be seen extended on the ground, examined by the eyes, and felt by the fingers. The specimen was a young female Rorqual, or razor-backed whale (so called from its having a fin on its back somewhat like a razor.) It was driven on shore at Winterton, eight miles from Great Yarmouth, Norfolk, in a tremendous gale from east-south-east, on the fifth of day of January, 1857. Her full length is forty-eight feet; her weight about twenty-five tons. The color of the skin is dark brown on the back, vanishing off towards the body in a bluish-grey. The tail measures, from tip to tip, eleven feet. This is composed of a dense fibrous mass, and feels to the touch like a thick sheet of india-rubber. It is placed at right angles to the body, in the reverse way to that usually seen in the fish. The eye is remarkably small, and the folds of the eyelids well marked: as it was impossible to preserve the eye in its natural bright state, an artificial glass model had been inserted into the eyelids, the natural colors of the eye having been closely imitated. The liver of this animal completely filled a one-horse cart, and was as much as the horse could draw. The heart about filled a good-sized washing-tub, and a section of the principal artery (the aorta) would about fit round an ordinary sized bucket. The weight of the blubber was not ascertained. It seems extraordinary that the captors were not aware of the value of the oil; for they cut the great masses of the blubber off, and spread it as manure over the fields. The fin which is placed by the side of the animal is remarkable; it contains four fingers, like human fingers, not, however, all separated one from the other, but inclosed in the skin of the fin, which looks like that of an ordinary fish. Imagine a human hand inserted into a hedge-cutter's glove, and wax poured round it, and you have (minus the thumb, of which the whale has no trace) an exact model of the whale's fin.

When the whale found itself on shore, it "roared loudly," as the proprietor described it. The noise was probably produced by the whale expelling air through his spiracles or blow-holes. A man went out into the water with an anchor, and rope attached, by way of a harpoon; twice, with all his force, did he dig the anchor into the fat blubber of the beast, twice did the beast by his tremendous struggles tear the weapon out again; but the third time, the anchor luckily turned, and thus caught about two feet of the skin in one of its flukes, and thus was the whale secured. The three gaping rents in the skin were plainly visible. The operator, however, had a dangerous task; for the whale, in its agonies, struck right and left with its tail, nearly drowning its enemy in the whirlpool caused thereby.

Its gigantic mouth is placed wide open by means of props, and a moderate-sized man can stand upright in it. This mouth was by far the most curious part of the exhibition; for in it can be seen, in their natural position, the plates of whalebone, or baleen, so much used, not only in the arts, but by ladies in almost every portion of their dress. Let the ladies consider how much they are indebted to the poor whale, as far as their dress is concerned; for how would they get out without whalebone? A fashion started by them extends its influence to distant regions of which they often have no idea. Before the invention of *crinoline*, the whales far away in the northern waters carried their baleen, or whalebone, in their mouths, and spouted water through it, thereby obtaining their dinners of minute sea animals in comparative peace and quietness. But, fashion invented crinoline. Whalebone was required to make it, and the price of whalebone went up from £150 per ton to £620. Beaver hats, and bonnets, and muffs have gone out of fashion; so the poor beavers have a rest, and are not nearly so much persecuted as they were in former days. It is now the whale's turn, and they are harpooned and otherwise slaughtered in order that their baleen, or whalebone, should be transferred from their mouths into ladies' dresses. No living creatures will be more pleased when the reign of crinoline is over than the whales; but many of them must yet fall victims to the fashion which has proved so "killing" to them.

Now, whalebone is by no means true bone. Put a bit of so-called whalebone by the side of the bone of a leg of mutton, and the difference will be perceived. There are three hundred and eighty plates on each side of the mouth; on the right side, the foremost hundred and twenty are of a beautiful milk-white, the rest being nearly black. This is simply a variety; some whales have been killed entirely white—they answer to the Albinos in the human species. Whalebone is composed of a substance of a horny appearance and consistence; internally it is of a fibrous texture, resembling hair; and the external surface consists of a smooth enamel, capable of receiving a good polish. It answers the purpose of teeth to the whale, and is placed in the position where teeth are usually found in other animals, in the upper jaw; none whatever are found in the lower jaw, which is covered by a hard, firm gum, as polished and as smooth as a mahogany table. Along each side of the jaw are found plates or layers of this whalebone. These can be counted from the outside, and look like the portions of a Venetian-blind when half opened; inside they cannot be counted, because they appear to be covered with hair. The hair is in reality nothing more than the actual substance of the baleen, unravelled, as it were, like tow from the end of a rope. If the reader wishes to prove this, let him take a thin bit of whalebone, boil it and soak it well, and then beat it with a wooden mallet. The result will be a bundle of coarse hair like horse-hair. This hair hangs in thick masses inside the beast's mouth; in the specimen I saw, it gave me exactly the idea of the long and beautiful silky white beard of a venerable old man. This is a thing which cannot be seen in any museum, and of which a picture would convey but an erroneous idea.

Upon going to the College of Surgeons, I found but few specimens of the baleen, but those very interesting. The indefatigable John Hunter dissected, among others, a bottle-nosed whale, which was cast ashore from the Thames in 1783. Its skeleton is now suspended from the roof of the new and magnificent room of the museum, and sections of its baleen are preserved in bottles. It appears, from his observations, that the baleen, like the teeth of rodent (or gnawing) animals, is endowed with perpetual growth, and that material is supplied from above as it is worn away from below; moreover, it is composed of three parts—the centre portion, being secreted from a soft cone, becomes hair—the external portions become horn, inclosing the hair. These three appear solid; but when the baleen has grown to a certain extent, the two external walls become worn off, and, as a matter of necessity, leave the hair exposed, so that, as said before, the mouth appears to be lined with hair. Aristotle has remarked this fact, for he writes: "The whale has hairs inside his mouth, in the place of teeth, like the bristles of a pig." A superficial observer, looking at our Whitechapel whale, would probably make exactly the same remark. In a picture I have of the Rorqual, there is drawn a tuft of hair projecting from the anterior end of the upper jaw. There is no real tuft there; but, upon examining the specimen, I perceived how the mistake originated. The baleen at this part consists entirely of hair, unconfined at either side by the side portions as above described. When the animal is in the water, this would probably float upward, giving the appearance of a tuft of hair on the tip of the nose.

It was aptly remarked by my lamented father, in one of his Oxford lectures, that the whale, being the largest of warm-blooded animals, and requiring a vast quantity of food to support its huge carcass, would have starved to death, if, like other creatures which have heart and lungs, and not gills like fish, it had been sent to sustain itself on land, either in the form of a carnivorous or gramnivorous animal. The Great Creator has in his omniscience, therefore, ordained that this, the largest of his creatures, should have the wide expanse of the ocean for his habitat; there, it would have plenty of room for its roamings, and plenty of food for its support. The whale, therefore, preserving every organ typical of the land animal, and remaining a true mammalian in every sense, associates with fishes, and grazes upon the products of the deep.

The sea, as we well know, swarms with life; but the minute creatures therein exceed by myriads the larger forms. Upon these atoms the whale feeds, and not only feeds, but gets fat, which fat it converts into blubber. Now, for the sake of this blubber, man will brave the perils of the Arctic seas, and bring home with him, in the form of valuable lamp-oil, the substance of acres of minute sea-creatures, which, but for this wise economy in the system of creation, would have lived and died neglected and useless. Thus we see, in the works of the benevolent Creator, wheel within wheel—nothing lost, nothing allowed to decay—all working together with an admirable and designed order. The creatures which principally form the food of the whale are a delicate mollusk called the Clio Borealis (of which specimens may be seen in the College of Surgeons). These creatures live in patches on the surface of the Northern Ocean; and could we look down on those Arctic Seas from a balloon, we should see greenish and blackish patches here and there; these are formed by colonies of the Clio Borealis. A somewhat similar appearance may be observed on stagnant fresh-water ponds, where the water is colored, here and there, by the larva of gnats and other insects.

Having found out the whereabouts of his food, the whale opens his gigantic mouth, and charges at full speed in among them; and I believe he has the power of actually smelling their whereabouts. Drawn into his mouth by the vast current of water created by the charge, like sticks in a mill-tail, they become engulphed in the natural trawl-net of the sea-giant, who then composedly shuts his mouth, and expels the water through the interstices of the baleen, leaving the Clios, and whatever else he is lucky enough to catch, high and dry upon the hairy roof of his mouth. In the specimen under notice, I observed that there were several folds of skin, extending from the tip of the lower jaw some distance down the belly; and the man informed me that when the lower jaw was lifted off the ground, the tongue was left on it some three feet below; the folds of skin at the same time becoming quite smooth. Here, then, we have an explanation of the use of these folds: they form an immense pouch, into which the detained animals drop, being freed from the hair. The bag of a lady's work-table gives a very good idea of the pouch of the whale —the silk portion representing the folds, and the board at the bottom the tongue.

The reader is not very likely ever to see a

whale at feed; he may, however, very likely see a duck feeding in a gutter. Let him observe, and he will see, that (to compare great thing with small) the duck goes to work in a very similar manner to the whale. The duck is looking after minute creatures—so is the whale; so he takes a billful of mud, and, squirting out the refuse, he retains what is good to eat. The bird has no baleen, and no pouch; but, nevertheless, he has an equally beautiful apparatus in the conformation of his bill, which answers the same purpose, and at the same time is less cumbersome. From the size of the whale's mouth, one would naturally be led to conclude that the gullet (or œsophagus) is of an enormous size. No such thing—it is exceedingly small. In the whale examined (forty-eight feet long), the entrance to the gullet is hardly large enough to admit a man's hand. Why is this? The Rorqual does not confine himself to the Clio Borealis; but he feeds upon sprats, herrings and little fish. If he had a capacious gullet, the fish having been swallowed might, not liking their new quarters, wish to return again to the sea; had the Whale an enormous gape, like a boa-constrictor, they might easily do this, as the stomach is on the same line as the mouth. This is, however, anticipated by the form to the œsophageal pipe. Upon examining a section of it, which is not much larger than the thickness of a good-sized walking-stick, we see that it has numerous muscular fibres surrounding it, and which can close it effectually; nay, more, the inner lining is disposed in longitudinal fibres the size of a little finger, which, meeting together in the centre, effectually render it impervious at the will of the animal.

Wishing to examine more minutely the base of the skull of the Whitechapel specimen, I crawled in through the place where the throat formerly was situated, and obtained an excellent view of the parts not externally visible.

At the College of Surgeons there is an enormous head of a whale (the bones only, without the baleen). It would contain three heads of the Whitechapel whale and an infinity of children. This was the first head ever seen in this country, and has been described and figured by the great Baron Cuvier himself. The form of the bones is that of three bows, two placed on the ground with their concave parts facing each other (the lower jaws), the third being represented by the upper jaw arching over them, its two ends corresponding with the points where the other bows touch one another.

I was exceedingly anxious to obtain the head of the Whitechapel whale, as a companion to the large head above mentioned, particularly as the whalebone, or baleen was in good condition. Knowing that the proprietor set great store on his acquisition, I approached the subject carefully, and was not surprised when he asked 125*l.* for his whale. It happened to be very warm weather just then, and when I was inside the whale's mouth I had observed that none of the bones of the head were in any way cleaned, or otherwise preserved, but still remained full of oil, &c. which, as a matter of course, I knew would soon become so offensive through the weather, that the proprietor would be only too glad to get rid of it at any price. It was only, therefore, necessary to bide one's time. My conjectures proved correct: in a week or so a letter came to Professor Quekett from the whale proprietor, offering to take less money; and as time advanced, and as the whale became more and more offensive, so did the price of the whale get less and less, the result being, that Professor Quekett, at the College one morning, received the whale's head, packed up in a large box, and sent back a cheque for 5*l*. only, instead of the 125*l*. originally demanded.

This head will shortly be articulated by Mr. Flower, and will be the only specimen in London showing the head of a large whale with the baleen *in situ*.

The whale skin is about as thick as six sheets of common note-paper placed together, and is very tough and elastic. It is covered with a scarf-skin. When dried, this thin scarf-skin peels off, and the true skin underneath is found to be of a jet black color, of a shining appearance, exactly like newly-tarred boards of a farm-yard fence that have been submitted to the heat of the sun.

Some years before I was born, a large whale was caught at the Nore, and towed up to London Bridge, the Lord Mayor having claimed it. When it had been at London Bridge some little time, the Government sent a notice to say the whale belonged to them. Upon which the Lord Mayor sent answer, "Well, if the whale belongs to you, I order you to remove it immediately from London Bridge." The whale was therefore towed down stream again to the Isles of Dogs, below Greenwich. The late Mr. Clift, the energetic and talented assistant of his great master, John Hunter, went down to see it. He found it on the shore, with its huge mouth propped open with poles. In his eagerness to examine the internal parts of the mouth, Mr. Clift stepped inside the mouth, between the lower jaws, where the tongue is situated. This tongue is a huge spongy mass, and being at that time exceedingly soft, from exposure to air, gave way like a bog, at the same time he slipped forwards towards the whale's gullet, nearly as far as he could go. Poor Mr. Clift was in a really dangerous predicament; he sank lower and lower into the substance of the tongue and gullet, till he nearly disappeared altogether. He was short in stature, and in a few seconds would, doubtless, have lost his life in the horrible oily mass, had not assistance been quickly afforded him. It was with great difficulty that a boat hook was put in requisition, and the good little man hauled out of the whale's tongue.

A few days after the visit of Mr. Murie and myself, the Gravesend whale was sold to Mr. Blaker, an oil and cart-grease manufacturer, who lives on the marsh on the other side of the river. We paid it two or three visits, and were allowed to take away all parts of the whale which were not required for making oil; the result was, that we obtained many most interesting and valuable preparations, which are now at the Royal College of Surgeons. While we were at work, the proprietor of Rosherville Gardens was bargaining for the skeleton, which he is *now* exhibiting in a beautiful state of preparation at the gardens

The whole scene fully realized the details of a very old print of "The Dissection of a Whale," thus quaintly described, "Cetus ingens quem Farœ Insulæ Incola ichthyophagi, tempestatibus appulsum securibus dissecant et partiuntur inter se." [A huge whale, which being cast up by a storm, the fish-eating inhabitants of the Faroe Islands dissect with axes, and divide among themselves.]

www.ingramcontent.com/pod-product-compliance
Lightning Source LLC
LaVergne TN
LVHW021137110826
845150LV00005B/1051

* 9 7 8 1 4 2 5 5 4 8 5 4 4 *